Causal Inference for Machine Learning Engineers

Durai Rajamanickam

Causal Inference for Machine Learning Engineers

A Practical Guide

 Springer

Durai Rajamanickam (ORCID)
University of Arkansas
Little Rock, USA

ISBN 978-3-031-99679-5   ISBN 978-3-031-99680-1 (eBook)
https://doi.org/10.1007/978-3-031-99680-1

This Springer imprint is published by the registered company Springer Nature Switzerland AG
The registered company address is: Gewerbestrasse 11, 6330 Cham, Switzerland

In learning you will teach,
And in teaching, you will learn.

—Phil Collins

Preface

In recent years, machine learning has revolutionized fields ranging from vision to language processing. However, most systems are designed primarily for pattern recognition, not for reasoning about cause and effect. This book, *Causal Inference for Machine Learning Engineers—A Practical Guide*, was born from the desire to bridge these two worlds. It offers a robust framework for answering critical questions about interventions, counterfactuals, and decision-making under uncertainty.

This guide aims to introduce readers to the core concepts of causal thinking, explain how to formulate causal questions, and demonstrate how deep learning architectures can be adapted for causal inference tasks. Written for students, researchers, and practitioners, it assumes minimal prior knowledge of causality or machine learning, combining intuitive explanations with mathematical foundations and practical examples. Through exercises and solutions provided at the end of each chapter, readers can reinforce their learning. My hope is that this book inspires readers to think causally and to explore the exciting new frontier where causality and deep learning meet.

Glastonbury, CT, USA   Durai Rajamanickam

Acknowledgments This book would not have been possible without many individuals' support, guidance, and encouragement.

First and foremost, I would like to express my gratitude to my mentors, my professors at UALR and colleagues who have shaped my understanding of causal inference and machine learning. Their questions, feedback, and insights constantly challenged me to think deeper and explain better.

I am especially thankful for the vibrant research community at the intersection of causality and deep learning. Their pioneering work laid the foundation for much of what is discussed in these pages.

My heartfelt gratitude goes to my family, including my wife, and my children, for their enduring patience and inspiration. I also extend my sincere thanks to my mentors, and my friends, for their unwavering belief in this project

Generative AI tools, including Google Gemini, ChatGPT, and Grammarly, were used to refine the language and improve the overall flow of the text, enhancing readability and presentation.

Furthermore, some examples employed throughout the text are intentionally drawn from concepts familiar to scholars and researchers, aiming to leverage existing knowledge for efficient learning.

Durai Rajamanickam

Competing Interests The author has no competing interests to declare that are relevant to the content of this manuscript.

Contents

About the Author

Durai Rajamanickam is an AI and Data Science leader with extensive experience in machine learning, causal inference, and real-world applications. His work spans healthcare, finance, legal analytics, and large-scale data engineering.

Driven by a deep passion for learning and teaching, he has spent years researching ways to make AI systems predictive and causally aware. He has led AI-driven business initiatives, and mentored students and professionals in advanced data science topics.

Durai believes that causal reasoning is essential for the next generation of intelligent systems and that bridging deep learning with causality will empower better decision-making in every field. This book reflects his journey toward making complex ideas accessible and fostering a new wave of causal thinkers in AI.

When he is not working on research or industry projects, Durai enjoys writing, mentoring, and exploring new ideas at the intersection of technology and human understanding.

Introduction to Causal Thinking

Understanding cause and effect is essential to making decisions, building explanations, and designing interventions. Whether in medicine, economics, social sciences, or artificial intelligence, causal thinking allows us to move beyond observing patterns and correlations to answering more profound questions: What causes what? What would happen if we intervened?

This chapter introduces the basic ideas of causal reasoning. We will explore why causality is different from correlation, why simple observational data often misleads us, and how formal models of causality (the study of cause and effect relationships, forming the foundation of causal inference [1]) can help. By building a solid foundation in causal thinking, we set the stage for more advanced techniques that combine traditional causal inference with the representational power of deep learning [2].

1 Rationale: Why Causality?

Prediction is not Causation. In modern data-driven fields, machine learning models excel at prediction tasks: given a dataset of inputs X and outcomes Y, models learn to approximate $P(Y \mid X)$ with remarkable accuracy. However, prediction alone cannot answer questions about **intervention**.

Consider a medical example: observing that individuals with higher physical activity levels have lower rates of cardiovascular disease enables us to predict health outcomes. Yet it does not tell us whether encouraging a sedentary individual to exercise will necessarily lower their disease risk.

© The Author(s), under exclusive license to Springer Nature Switzerland AG 2025

D. Rajamanickam, *Causal Inference for Machine Learning Engineers*,

https://doi.org/10.1007/978-3-031-99680-1_1

This distinction arises because hidden variables may confound observed associations. We require a causal framework to answer interventional questions, such as "What would happen if we changed X?"

At its heart, causal inference is about understanding the consequences of actions. It asks: not merely *what is*, but *what would happen if*.

Key Distinction **Prediction:** Given X, what is the likely Y?
Causation: If we set X to a specific value, what is the resulting Y?

In formal terms, prediction answers questions about $P(Y \mid X)$. At the same time, causality demands knowledge of $P(Y \mid do(X))$, where the *do-operator $do(X = x)$* denotes an active intervention.

These quantities are generally *not the same*:

$$P(Y \mid X) \neq P(Y \mid do(X))$$

Unless strong assumptions, such as no confounding, hold.

Distinguishing Correlation from Causation is foundational across science, engineering, and medicine policy.

1.1 Intervention: The Defining Query

Suppose you are tasked with answering:

"Will a new job training program improve employment rates?"

Predictive models trained on past data may indicate that people who underwent training had higher employment. However, such models cannot distinguish whether the training caused the improvement or whether those who chose to participate were inherently more motivated.

Any intervention based on predictive models risks being ineffective or harmful without causal reasoning.

Causal and ML

Machine Learning in Practice

Causal models often generalize better to new, unseen data distributions, a key challenge in machine learning.

- **Prediction versus Control:** Machine learning models are like skilled predictors; they spot patterns to forecast outcomes. But causal inference is about being a **controller**; it helps us predict what happens if we change something.
- **Causality as a Robust Compass:** Think of a weather forecast (prediction) versus understanding how air pressure changes cause storms (causality). The causal understanding is more reliable if the climate changes. Similarly, causal models tend to be more robust when the data changes.
- **Feature Selection Analogy:** Feature selection in machine learning is like choosing ingredients for a recipe. Causal inference helps us pick the **active** ingredients (causes) rather than just those that happen to be in the pantry (correlations).
- **Example (Recommendation Systems):**

    - Prediction: The model sees "Users who browsed X also bought Y" and recommends Y.
    - Causation: Causal inference asks, "Does recommending X **actually** make them buy Y?" Maybe they would have bought it anyway!
    - Causal recommendations are like giving users what they **need**, not just what they were going to buy.

1.2 Real-World Motivations

Healthcare: Estimating the effect of a new drug on patient survival.
Economics: Determining whether education subsidies increase earnings.
Social Sciences: Evaluating the impact of minimum wage laws on employment.
Machine Learning: Building robust models under distributional shifts capable of counterfactual reasoning.

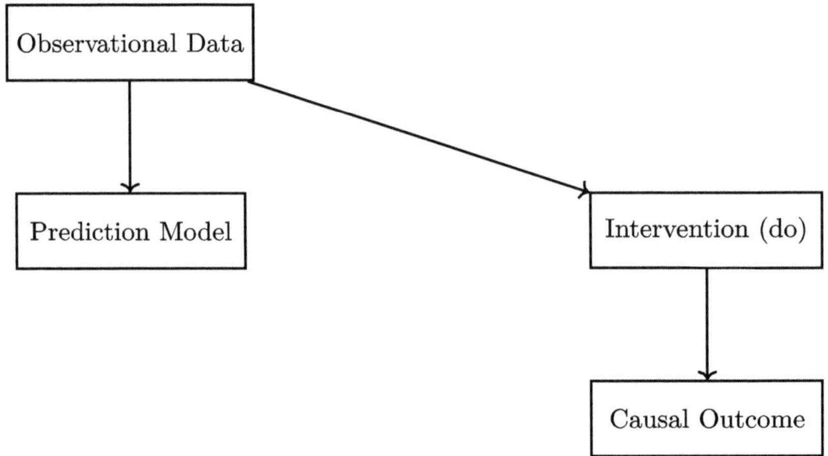

Fig. 1 Visualizing prediction versus causation

1.3 A Graphical Intuition

The differences between prediction and causation can be visualized (Fig. 1).
Interpretation: - Prediction relies on associations within the observational data. - Causal inference requires modeling the outcome of interventions that may not be present in the data.

2 What Is Causality?

Causality is the study of cause-and-effect relationships. It seeks to understand how changes in one variable bring about changes in another. In contrast to purely associational learning, which identifies patterns within data, causal reasoning aspires to uncover the underlying mechanisms that govern these patterns.

To speak of causality is to move beyond observing that two variables X and Y are related. It is to assert that changes in X *cause* changes in Y.

Formally, if X and Y are two random variables, we say that X causes Y if intervening to set X to different values changes the distribution of Y.

This intuition is captured mathematically via the **do-operator** introduced by Judea Pearl:

$$P(Y \mid do(X = x))$$

which denotes the probability distribution of Y under an intervention that sets X to x.

The key difference between observing $X = x$ and actively setting $X = x$ is that interventions break the natural causal pathways leading to X. They eliminate the influence of confounding variables that may otherwise affect both X and Y.

2.1 An Example: Aspirin and Headache Relief

Suppose we observe that individuals who take Aspirin tend to recover from headaches faster.

Is it safe to conclude that Aspirin causes headache relief?

Merely observing $P(\text{Relief} \mid \text{Aspirin})$ may be misleading. Perhaps individuals who take aspirin are more health-conscious overall, and their general health practices—rather than the aspirin itself—explain the recovery.

To infer causality, we would ideally randomize the aspirin assignment: giving it to some individuals regardless of their natural tendencies. This intervention would allow us to estimate $P(\text{Relief} \mid do(\text{Aspirin}))$, and thereby identify the true causal effect.

Causal and ML

Causality for Machine Learning Practitioners

Adversarial examples in machine learning highlight the danger of relying on non-causal features. Causal models are inherently more resilient to such attacks.

- **Beyond the Black Box:** Machine learning models are often "black boxes." Causal inference helps us open them up and see **why they make decisions, not just what they predict**.
- **Causality as Stability:** Imagine a self-driving car. A purely predictive model might slam on the brakes whenever it sees a shadow. A causal model understands that shadows don't **cause** collisions and is more stable.
- **Example (Image Recognition):**

  - Prediction: The model says, "This is a bird picture."
  - Causation: Causal inference asks, **"What parts of the picture—the beak, the wings—made the model recognize a bird?"**
  - A causal model focuses on essential features, like a detective looking for real clues.

2.2 Causality in Three Key Questions

Causal inference typically revolves around three key questions:

1. **Prediction (Associational):** What is $P(Y \mid X = x)$?
   Example: Among aspirin patients, what proportion recover quickly?
2. **Intervention (Causal Effect):** What is $P(Y \mid do(X = x))$?
   Example: If we administer aspirin, what will be the chance of recovery?
3. **Counterfactual (Retrospective):** What would have happened if X had been different?
   Example: For a patient who took aspirin and recovered, what would have happened had they not taken aspirin?

Each of these questions is progressively harder and requires stronger assumptions.

2.3 Formalizing Causal Models

The modern theory of causality is built upon two major frameworks:

- **Structural Causal Models (SCMs):** [1] A system of equations representing how each variable is generated from others and possibly some random noise.
- **Potential Outcomes Framework:** A counterfactual model introduced by Neyman and Rubin, where each unit has multiple potential outcomes depending on the treatment received

  In Structural Causal Models, variables are generated according to structural assignments:
$$Y = f(X, U_Y)$$

where U_Y is exogenous noise, independent of X, intervening on X replaces its structural assignment, disconnecting it from its usual causes.

2.4 Causal Graphs: A Visual Representation

Causal relations are often visualized using directed acyclic graphs (DAGs), where: - Nodes represent variables. - Directed edges ($X \rightarrow Y$) indicate direct causal influence (Fig. 2).
   For example, in the aspirin scenario:

Interpretation: Health consciousness affects both the likeli- hood of taking aspirin and the likelihood of recovery. - Aspirin may have a direct causal effect on relief. - Confounding must be accounted for to isolate the true causal effect of aspirin.

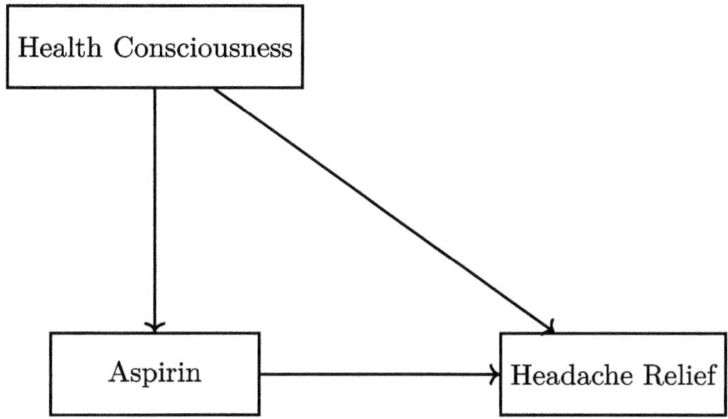

Fig. 2 Causal graph visual representation

2.5 Why Causal Thinking Matters

Without causal understanding:

- **Predictions may fail under interventions. - Policies may be based on spurious correlations. - Machine learning models may generalize poorly when the environment shifts**.

Causal models are essential for **counterfactual reasoning, decision-making**, and **robustness** across domains.

Thus, causality is not merely an academic concern; it is at the foundation of responsible and effective data-driven practice.

3 Correlation and Causation

One of the most important and critical lessons in scientific reasoning and statistical learning is that correlation does not imply causation. Two variables may be statistically associated without one causing the other. Understanding why this is so requires carefully analyzing the sources of the observed associations.

> **Causal and ML**
>
> ### Why Correlation Fails in Machine Learning
>
> Machine learning models, by default, optimize for prediction accuracy on the training data, which can lead to the exploitation of spurious correlations if not addressed.
>
> - **The Problem of Spurious Correlations:** Machine learning models are like computers that are very good at spotting patterns. But they can be tricked by "glitches" in the data—spurious correlations.
> - **Causality as a Fix:** Causal inference is like a debugging tool that helps us find and fix those glitches, making our models more reliable.
> - **Fairness Analogy:** Imagine a hiring algorithm. If it sees that people with certain names are more likely to get hired, it might unfairly favor those names. Causal reasoning helps us remove these unfair "causes."
> - **Example (Loan Approval):**
>   - Prediction: The model notices "People in this neighborhood are more likely to repay loans."
>   - Causation: Causal inference asks, "What **really** makes someone likely to repay? Their job? Their income?"
>   - Causal models aim for fair decisions, based on real reasons.

3.1 The Danger of Naive Interpretation

Suppose we observe a strong correlation between the number of firefighters at a fire and the amount of property damage incurred. Would it be reasonable to conclude that firefighters cause more damage? No. The causal explanation is that larger fires cause more property damage and require more firefighters. The correlation arises from a *common cause*, not a direct causal effect.

Similarly, a famous example relates to ice cream sales and drowning incidents:
- Ice cream consumption and swimming activity increase on hot summer days. - Consequently, drowning incidents and ice cream sales are correlated. - But ice cream consumption does not cause drowning.

These examples illustrate that **correlation is insufficient to infer causality**.

3.2 Mathematical Formalization

Two variables X and Y are said to be **correlated** if their joint distribution differs from the product of their marginals:

$$P(X, Y) \neq P(X)P(Y)$$

However, Correlation measures symmetric statistical dependence:

$$\mathrm{Corr}(X, Y) = \mathrm{Corr}(Y, X)$$

While causality is inherently asymmetric:

$$X \to Y \quad \text{but not necessarily} \quad Y \to X$$

Thus, correlations can arise from:

1. **Direct causation:** $X \to Y$
2. **Reverse causation:** $Y \to X$
3. **Confounding:** $Z \to X, Z \to Y$
4. **Coincidence:** Random statistical flukes in finite samples.

3.3 Visualizing Correlation Versus Causation

Consider the causal graph corresponding to the firefighter example.

In Fig. 3, **Fire Severity** is the common cause of the number of firefighters and the extent of damage.

The observed Correlation between *Firefighters* and *Damage* is spurious unless we account for the confounder *Fire Severity*.

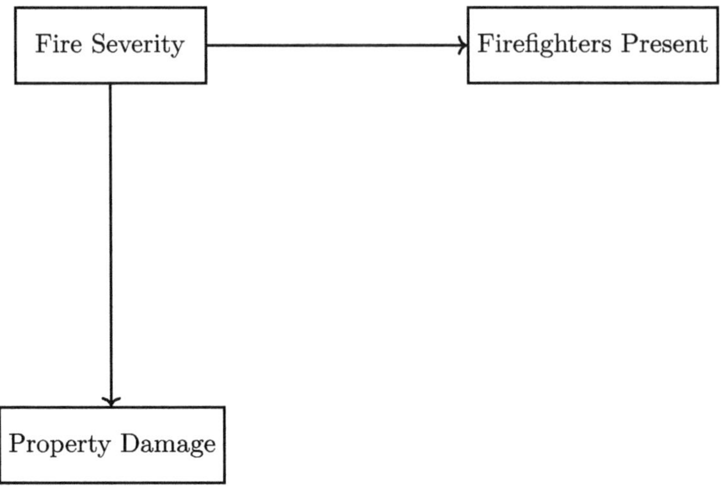

Fig. 3 Correlation versus causation

Causal and ML

3.4 Relevance to Machine Learning

Standard machine learning models, including deep learning systems, learn from correlations in the training data. Without explicit causal reasoning:

- **Predictions** may exploit superficial patterns that fail under distribution shifts.
- **Policy** recommendations based on correlations may have harmful consequences.
- **Generalization** outside the training distribution remains unreliable.

For example, if a healthcare model correlates treatment recovery without accounting for patient severity, its deployment may harm sicker patients.

Thus, causal understanding is philosophically essential and practically indispensable for building robust, reliable systems.

3.5 Towards Causal Learning

To move from correlation to Causation, we require either:

- **Interventions:** Actively manipulating variables and observing effects (e.g., randomized controlled trials).

- **Assumptions and Models:** Structural causal models or counterfactual frameworks that allow causal identification from observational data.

4 Causal Questions We Seek to Answer

Causal inference is ultimately about answering questions that go beyond mere observation. Instead of asking *what is happening*, causal analysis asks *what would happen if we acted differently*.

This shift requires a precise formulation of the types of causal questions we aim to address.

Causal and ML

The Ladder of Causal Questions for ML

- **ML's Comfort Zone:** Standard machine learning lives in the "Association" zone, finding patterns.
- **Stepping Up the Ladder:** Causal inference helps us climb higher:

  - Intervention: "What happens if we **do** this?" (like reinforcement learning).
  - Counterfactuals: "What **would have** happened if...?" (like explaining a decision).

- **Counterfactuals for Decision-Making:** Causal ML is like giving our models a powerful "What if?" button so that they can reason about different scenarios.
- **Example (Autonomous Driving):**

  - Association: The car sees "When it rains, there are more accidents."
  - Intervention: The car asks, "If I **slow down, will it actually** reduce the risk of an accident?"
  - Counterfactual: The car wonders, "If I had **not** changed lanes, would that accident have happened?"
  - Causal reasoning helps the car make **smart** choices.

4.1 The Three Key Questions

Counterfactual reasoning is crucial for building machine learning systems that can explain their decisions and assess fairness.

According to Judea Pearl's seminal framework [1], causal reasoning revolves around three key questions:

1. **Association:** What can we predict?
2. **Intervention:** What if we do something?
3. **Counterfactual:** What would have happened if things were different?

These questions form the **Ladder of Causation**, [1] which we will study in depth later. Here, we introduce them informally.

Association (Seeing) Given an observation X, what is the probability of another variable Y?

$$P(Y \mid X)$$

Example:

- Given that a patient has high blood pressure (X), what is the likelihood they will experience a stroke (Y)?

This is the domain of conventional statistical and machine learning methods.

Intervention (Doing) What is the effect of actively setting X to a value x?

$$P(Y \mid do(X = x))$$

Example:

- If we prescribe a blood pressure medication $(X = 1)$, what is the probability of reducing the risk of stroke (Y)?

The **do-operator** distinguishes intervention from observation. This requires causal reasoning beyond statistical associations.

Counterfactual (Imagining) What would have happened to Y if, contrary to fact, X had been different?

$$P(Y_{x'} \mid X = x, Y = y)$$

Example:

- For a patient who did not receive the medication and suffered a stroke, what is the probability they would have avoided the stroke had they taken the medication?

Counterfactuals are the most subtle and complex causal questions, underlying domains like law, ethics, and personalized decision-making.

4.2 *Illustrative Example: School Program and Student Success*

Suppose a school implements a tutoring program aimed at improving student test scores.

We might ask:

- **Association:** Association: Among students who enrolled in the program, how many had high scores? This is represented by $P(\text{High Score} \mid \text{Tutoring} = 1)$.
- **Intervention:** Intervention: If we assigned the tutoring program to all students, what would happen to their scores? This is represented by $P(\text{High Score} \mid do(\text{Tutoring} = 1))$.
- **Counterfactual:** For a specific student who did not attend tutoring and failed, would they have passed if they had attended?

4.3 *From Prediction to Causal Understanding*

Traditional machine learning focuses on association: given past data, predict future outcomes (Fig. 4).

Causal inference enables us to:

- Predict the effects of new policies or treatments not present in historical data.
- Understand the mechanisms underlying observed phenomena.
- Reason about alternate realities to support better decision-making.

This ability is indispensable in fields such as:

- Healthcare: Which treatments cause better outcomes?
- Economics: What is the impact of policy changes on employment?
- Education: How do interventions affect student success?
- Social Sciences: How does social media usage influence mental health?

Understanding causal questions requires moving beyond data fitting to modeling the process that generates the data. By the end of this book, you will be equipped to address each type of causal query systematically—and to distinguish when data alone is insufficient to answer them.

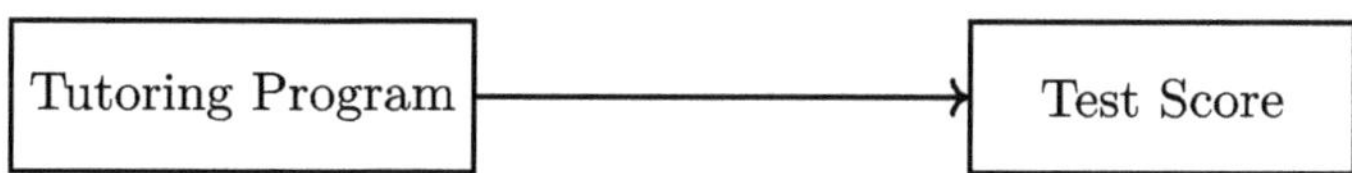

Fig. 4 Tutoring and testscore association

5 The Ladder of Causation

One of the most profound contributions to modern causal inference is Judea Pearl's [1] **Ladder of Causation**. It classifies causal reasoning into three levels, each building upon the previous.

Understanding this ladder is essential because it defines what can be learned from data, and what requires causal assumptions or interventions.

5.1 The Three Rungs of the Ladder

1. **Association (Seeing)**: Detecting statistical patterns in data.

$$P(Y \mid X)$$

2. **Intervention (Doing)**: Predicting the effects of actions.

$$P(Y \mid do(X))$$

3. **Counterfactuals (Imagining)**: Reasoning about alternate realities.

$$P(Y_{x'} \mid X = x, Y = y)$$

5.2 Visual Representation of the Ladder

Each level answers a different class of questions and demands stronger causal assumptions and richer data representations (Fig. 5).

5.3 Rung 1: Association (Seeing)

Definition: Observing statistical correlations between variables.

$$P(Y \mid X)$$

Examples:

- How likely is a patient to recover given that they took a medication?
- How frequently do two words co-occur in a text corpus?

Fig. 5 The ladder of causation

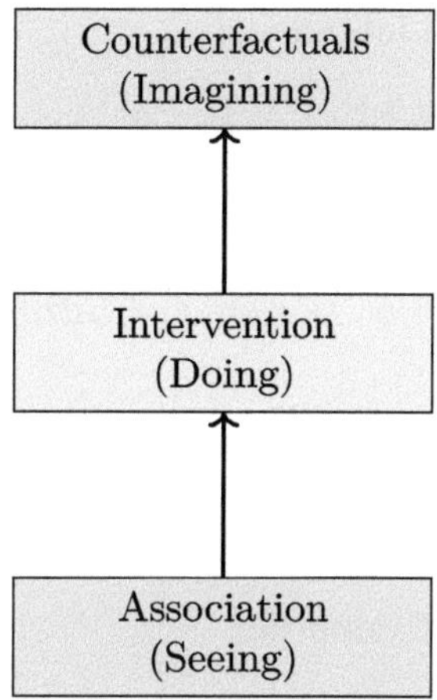

Capabilities:

- Purely observational analysis.
- Achievable using standard supervised learning algorithms.

Limitations:

- Cannot predict effects of actions or interventions.
- Confounding can create spurious associations.

5.4 Rung 2: Intervention (Doing)

Definition: Predicting the effects of deliberately setting a variable.

$$P(Y \mid do(X))$$

Examples:

- What is the effect on blood pressure if we administer a specific drug?
- How would banning ads on social media platforms affect the spread of misinformation?

Capabilities:

- Requires the concept of manipulating variables.
- Achievable using causal models like causal graphs and do-calculus [1].

Limitations:

- Requires assumptions about the data-generating process.
- Randomized controlled trials (RCTs) or good causal models are needed.

5.5 *Rung 3: Counterfactuals (Imagining)*

Definition: Reasoning about what would have happened under alternate circumstances.

$$P(Y_{x'} \mid X = x, Y = y)$$

Examples:

- Would a cancer patient have survived if they had received chemotherapy instead of surgery?
- Would a rejected loan applicant have been approved had they slightly increased their income?

Capabilities:

- Enables personalized decision making.
- Forms the foundation of fairness, explainability, and ethical AI.
- Powers retrospective analyses (e.g., legal liability cases).

Limitations:

- Requires full causal models that encode structural relationships.
- Observational data alone is insufficient.

5.6 *Why the Ladder Matters*

Many machine learning systems operate only at the bottom rung, extracting patterns without understanding cause and effect.

In contrast, causal reasoning:

- Allows us to predict the consequences of new actions.
- Helps control for confounding and selection bias.
- Enables deeper reasoning about "what could have been."
- Is essential for generalization beyond the training distribution (out-of-distribution).

5.7   Practical Implications

Consider building a recommendation system:

- **Association:** Recommend items similar to past clicks.
- **Intervention:** Estimate whether recommending an item will increase user engagement.
- **Counterfactual:** For a user who did not buy a product, infer whether they would have bought it if shown different recommendations.

Only with causal reasoning (rungs 2 and 3) can we build robust systems that adapt and reason under change.

The ladder of causality defines the scope of what is possible from data. To reach higher rungs, we must go beyond observational data and incorporate causal assumptions, models, and reasoning. Throughout this book, we progressively climb the ladder, equipping ourselves to answer more profound and practical questions about the world.

6   Challenges of Observational Data

Causal inference is more complex than predictive modeling because the data we collect is often *observational* rather than *experimental*.

Observational data reflect correlations shaped by many hidden forces, but not necessarily causal effects. Understanding the pitfalls of observational data is crucial before we embark on building causal models.

6.1   Confounding Variables

A **confounder** is a variable that influences both the treatment and the outcome, creating a spurious association (Fig. 6).

Example:

- **Treatment (X):** Taking vitamin supplements.
- **Outcome (Y):** Living longer.
- **Confounder (Z):** Health-consciousness: People who take vitamins might also exercise and eat healthy, confounding the apparent effect.

Impact: If we ignore Z, we might incorrectly conclude that taking vitamins causes a more extended life.

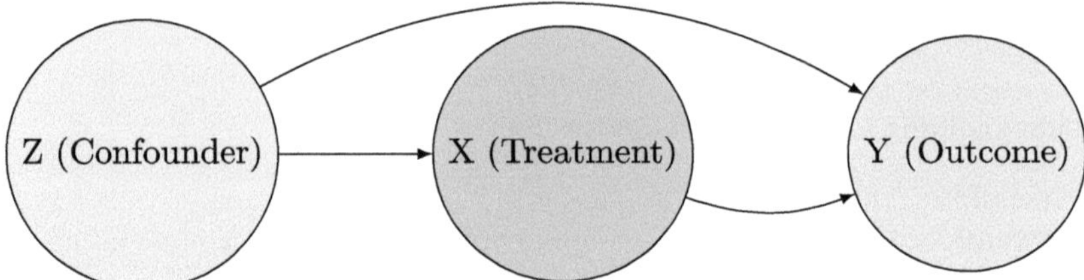

Fig. 6 Confounding variables

6.2 Selection Bias

Selection bias occurs when the way data is collected depends on the variables of interest (Fig. 7).

Example:

- Studying the effect of a new drug based only on patients who volunteer for a clinical trial.

Impact: The volunteers might differ systematically from the general population (e.g., healthier, wealthier), biasing the estimated treatment effect: Treatment Effect.

6.3 Simpson's Paradox

Sometimes, trends observed within subgroups reverse when the groups are combined.

Example: Consider university admission rates:

Fig. 7 Selection bias

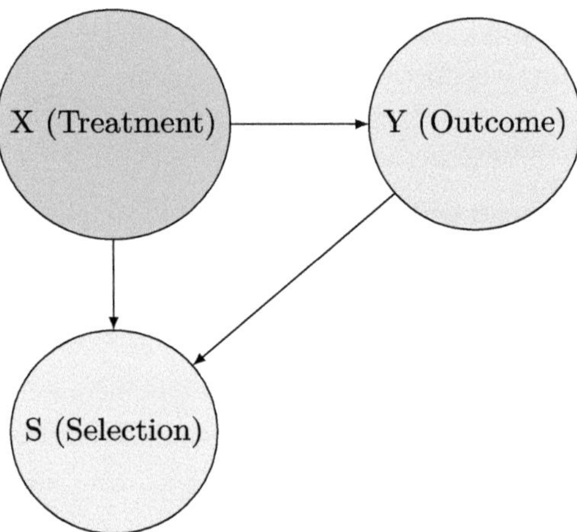

- Department A admits 80% of women, 70% of men.
- Department B admits 30% of women, 20% of men.

Yet overall, women have a lower admission rate because they apply more often to the more competitive Department B.

Impact: Aggregating data can hide or reverse causal relationships.

6.4 Missing Data

Missing data is pervasive in observational studies:

- Patients lost to follow-up.
- Unrecorded variables like genetic predisposition or past trauma.

Naive analyses become biased if the missingness is related to treatment or outcome.

6.5 Confounding Versus Selection Bias Versus Missing Data

Issue	Definition	Consequence
Confounding	Hidden common causes	Spurious correlations
Selection Bias	Biased sampling	Non-representative samples
Missing Data	Unobserved values	Incomplete information

Each of these issues demands different strategies to correct.

6.6 Why Observational Data Alone Is Insufficient

When data is purely observational:

- We cannot distinguish cause from mere correlation without assumptions.
- Observed dependencies may be distorted by confounding, selection, or missingness.
- Predictive accuracy does not imply causal validity.

Illustrative Example: A machine learning model trained to predict who buys insurance may exploit correlations between ZIP code and income, but this does not mean living in a particular ZIP code causes insurance purchases.

Without causal reasoning, interventions based on such models can backfire.

6.7 The Road Ahead

In later chapters, we will systematically develop tools to address these challenges:

- Using causal graphs to identify confounders.
- Applying do-calculus [1] to separate association from Causation.
- Designing algorithms that explicitly correct for biases.

Observational data is abundant but treacherous. The dangers of confounding, selection bias, and missing data make causal inference an art that demands careful modeling. Recognizing these challenges is the first step toward building systems that reason not only about what *is*, but also about what *could be*.

7 How Causal Inference Works

Causal inference is answering questions about cause and effect based on data, assumptions, and reasoning.

Unlike pure machine learning, causal inference seeks to understand not just the *what*, but the *why* and *what if*.

7.1 The Building Blocks of Causal Inference

At its core, causal inference rests on three pillars:

- **Assumptions:** All causal conclusions depend on assumptions about the data-generating process. These assumptions must be explicit and justifiable.
- **Modeling:** Representing the causal structure (e.g., via graphs or structural equations) enables us to formalize our assumptions.
- **Identification:** Using the model, we derive expressions for causal quantities purely from observed data.
- **Estimation:** We compute numerical estimates from finite samples, using statistical and machine learning techniques.

This logical pipeline distinguishes causal inference from purely predictive data analysis (Fig. 8).

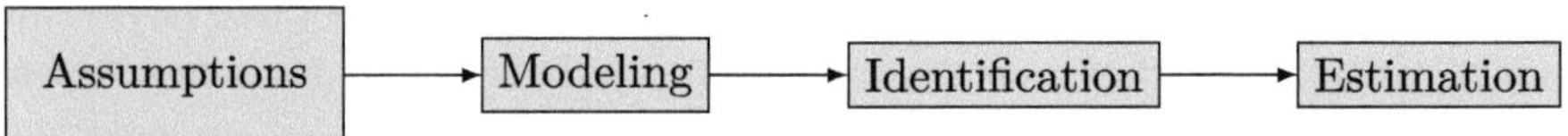

Fig. 8 Building blocks of causal inference

7.2 *Structural Causal Models (SCMs)*

A Structural Causal Model [1] formalizes causal assumptions using:

- **Variables** representing observed and unobserved quantities.
- **Structural Equations** describing how each variable is generated as a function of others.
- **Causal Graphs** visualizing the system of equations as a directed acyclic graph (DAG).

Definition: An SCM $\mathcal{M}$ is a 3-tuple (U, V, F) where:

- U = set of unobserved variables (noise terms).
- V = set of observed variables.
- F = set of structural functions, one for each V_i, mapping parents and noise to V_i.

Example:

$$U_X \sim \mathcal{N}(0, 1)$$
$$U_Y \sim \mathcal{N}(0, 1)$$
$$X = U_X$$
$$Y = 2X + U_Y$$

The causal graph.
In Fig. 9, X causes Y.

7.3 *The Language of Interventions*

In causal inference, we distinguish between:

- **Observation:** Passive seeing, like $P(Y|X = x)$.
- **Intervention:** Actively setting X to x, written as $P(Y \mid \mathrm{do}(X = x))$.

The **do-operator** $\mathrm{do}(X = x)$ signifies intervention, severing X from its natural causes.

Example:

$$P(\text{Cancer} \mid \mathrm{do}(\text{Smoking} = 1)) \quad \text{versus} \quad P(\text{Cancer} \mid \text{Smoking} = 1)$$

The first asks: "If we force people to smoke, what happens?" The second just passively observes smokers, who may differ from non-smokers in many ways.

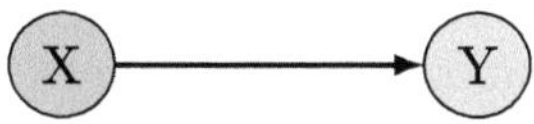

Fig. 9 Correlation due to a common cause (confounding)

7.4   Identification: Bridging Assumptions and Data

Given a causal query (e.g., $P(Y \mid \mathrm{do}(X))$), we need to express it in terms of observed data.

Identification uses:

- **Causal Graph Rules** (e.g., backdoor criterion [1], frontdoor criterion [1]).
- **Do-Calculus** (rules of manipulating do expressions).

Backdoor Criterion: If a set of variables Z blocks all backdoor paths from X to Y, then:

$$P(Y \mid \mathrm{do}(X)) = \sum_z P(Y \mid X, Z = z) P(Z = z)$$

Thus, causal effects can be computed by adjusting for Z.

7.5   Estimation: Computing Quantities from Data

Once identification yields an estimable formula, we estimate it using techniques such as:

- Regression adjustment.
- Propensity score methods.
- Matching and weighting.
- Deep learning models for counterfactual estimation.

7.6   An Overview in Equations

Suppose:

$$\text{Causal Query:} \quad P(Y \mid \mathrm{do}(X))$$

$$\text{Identification:} \quad P(Y \mid \mathrm{do}(X)) = \sum_z P(Y \mid X, Z = z) P(Z = z)$$

$$\text{Estimation:} \quad \text{Fit models for } P(Y \mid X, Z) \text{ and } P(Z)$$

Thus, causal inference links assumptions, models, and estimation seamlessly (Fig. 10).

Causal inference works by modeling assumptions about how variables interact, deriving quantities of interest using logical rules, and estimating these quantities from data. Unlike predictive modeling, causal modeling is an *assumption-driven science*—where assumptions enable conclusions.

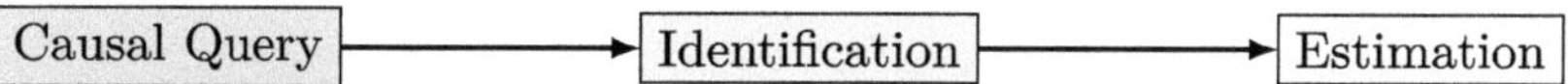

Fig. 10 Causal query model

8 Applications of Causal Inference

Causal inference is not confined to academic curiosity; it lies at the heart of decision-making across various disciplines. Understanding cause and effect allows practitioners to design better interventions, policies, and systems.

In this section, we survey major domains where causal thinking is crucial.

8.1 Healthcare and Medicine

In healthcare, causal inference answers critical questions directly impacting patients' lives.

Example Questions:

- Does a new drug *cause* recovery faster than existing treatments?
- What is the effect of early diagnosis on disease progression?
- How does vaccination *reduce* infection rates?

Case Study: Randomized Controlled Trials (RCTs) are the gold standard for causal inference in medicine. Patients are randomly assigned to treatment or control groups to ensure that outcome differences are causally attributable to the treatment.

Causal Graph Representation: In observational studies where RCTs are infeasible, adjusting for confounders like age, comorbidities, and socioeconomic status is essential (Fig. 11).

Fig. 11 Causal graph

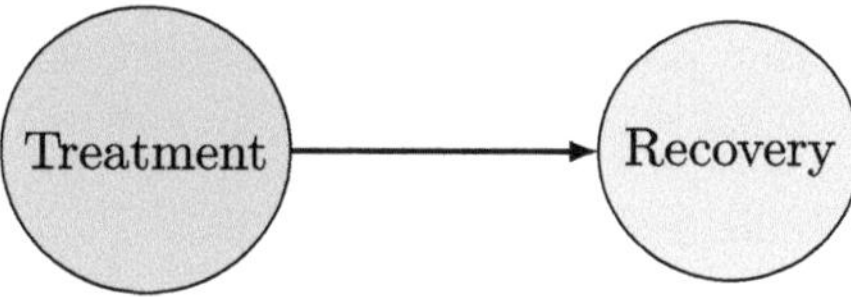

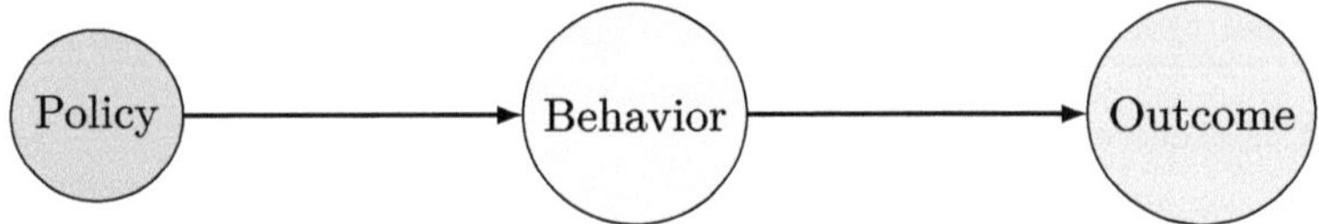

Fig. 12 Policy intervention

8.2 Economics and Policy Making

Governments often ask causal questions about policies before implementation.

Example Questions:

- What is the effect of a minimum wage increase on employment?
- Does providing subsidies *cause* increased college enrollment?
- How does taxation affect investment behaviors?

Econometric Techniques: Methods like Instrumental Variables (IV), Difference-in-Differences (DiD), and Regression Discontinuity Design (RDD) stem from causal reasoning, formalized long before machine learning became popular.

Diagram: Policy Intervention Causal graphs help separate direct and indirect effects of policies (Fig. 12).

8.3 Social Sciences

In psychology, sociology, and education, causal questions often involve complex human behavior.

Example Questions:

- Does parental involvement *cause* better academic outcomes?
- How does media exposure affect voting behavior?
- What is the impact of early childhood programs on long-term income?

Challenges:

- Observational data dominate.
- Confounding is pervasive.
- Ethical constraints limit experimental interventions.

Causal inference tools allow researchers to estimate effects despite these challenges by modeling assumptions carefully.

8.4 Technology and Machine Learning

Modern machine learning increasingly integrates causal reasoning.

Applications:

- Recommendation Systems: How does showing a product influence purchases?
- Advertising: What is the causal effect of an ad on customer conversion?
- Fairness: Does an algorithm cause disparate impacts on different groups?

Why Causality Matters in ML:

- **Generalization:** Causal relationships are more stable across environments than correlations.
- **Robustness:** Causal models are less sensitive to distribution shifts.
- **Interpretability:** Understanding *why* a model recommends an action is critical in sensitive domains.

8.5 Public Health and Epidemiology

During pandemics or public health crises, causal questions become urgent.

Example Questions:

- How effective are masks and social distancing in reducing transmission?
- Does early lockdown reduce mortality?

Real-World Challenge: During COVID-19, many studies tried to estimate the causal effect of interventions from non-randomized, messy, and confounded data.

Causal diagrams guided which biases needed adjustment, and which effects were identifiable.

9 Chapter Summary

Across medicine, economics, social science, technology, and public health, causal inference offers a structured approach to understanding and answering "what if" questions essential for improving human life and society.

Causal reasoning extends beyond correlations, enabling actionable, reliable, and interpretable decisions across disciplines.

10   A Roadmap for This Book

This book guides you from the importance of causal thinking to the frontier of causal machine learning and causal deep learning methods, with a structured and logical progression. Each chapter builds upon concepts introduced earlier, ensuring a coherent and cumulative learning experience.

11   Code

Listing 1.1   Chapter 1: Correlation and Causation

```python
import numpy as np
import pandas as pd
import matplotlib.pyplot as plt
import statsmodels.api as sm   # For regression analysis
    (optional)

# Simulate data for the ice cream sales and drowning example
np.random.seed(0)   # for reproducibility
n = 1000   # number of data points

# Temperature is the common cause (confounder)
temperature = np.random.normal(25, 5, n)   # average
    temperature of 25 degrees Celsius with some variation

# Ice cream sales increase with temperature
ice_cream_sales = 10 + 2 * temperature + np.random.normal(0,
    10, n)   # base sales + temp effect + noise

# Drowning incidents also increase with temperature
drowning_incidents = 5 + 1.5 * temperature + np.random.
    normal(0, 5, n)   # base incidents + temp effect + noise

# Create a Pandas DataFrame to hold the data
data = pd.DataFrame({
    'Temperature': temperature,
    'IceCreamSales': ice_cream_sales,
    'DrowningIncidents': drowning_incidents
})

# Calculate the correlation between ice cream sales and
    drowning incidents
correlation = data['IceCreamSales'].corr(data['
    DrowningIncidents'])
print(f"Correlation between ice cream sales and drowning:
    {correlation:.2f}")

# Visualize the data
```

```python
plt.scatter(data['IceCreamSales'], data['DrowningIncidents'
    ])
plt.xlabel('Ice Cream Sales')
plt.ylabel('Drowning Incidents')
plt.title('Correlation vs. Causation')
plt.show()

# --- Optional: Demonstrate with Regression ---
# Fit a simple linear regression model to predict drowning
    incidents from ice cream sales
model = sm.OLS(data['DrowningIncidents'], sm.add_constant(
    data['IceCreamSales'])).fit()
print(model.summary())  # Print regression summary

# Explanation of the code:
# 1. Simulation:
#    - We generate 'temperature' as the common cause.
#    - 'ice_cream_sales' and 'drowning_incidents' are both
    made to depend on 'temperature',
#      creating a correlation between them.
#    - Noise is added to each to make it more realistic.
# 2. DataFrame:
#    - Pandas is used to organize the data, making it easier
    to work with.
# 3. Correlation:
#    - The '.corr()' method calculates the Pearson
    correlation coefficient.
#    - A high correlation will be observed.
# 4. Visualization:
#    - A scatter plot visually shows the relationship. The
    plot will show that as ice cream sales increase, so do
    drowning incidents.
# 5. Regression (Optional):
#    - statsmodels is used to fit a linear regression.
#    - The regression output provides more detail (e.g.,
    statistical significance), further emphasizing the
    apparent relationship.
#
# Key takeaway: The code simulates a scenario where a
    correlation exists due to a common cause.
# The regression model might suggest a predictive
    relationship, but it doesn't mean ice cream sales cause
    drowning.
```

12 Exercises

1. Define causality in your own words. How is it different from correlation?
2. Give a real-world example where confusing correlation with causation could lead to a wrong decision.

3. What is the main goal of causal inference?
4. Describe an everyday situation where counterfactual thinking helps you make better choices.

References

1. Pearl, Judea. 2009. *Causality: Models, Reasoning, and Inference*. 2nd ed. Cambridge University Press.
2. Goodfellow, Ian, Yoshua Bengio, and Aaron Courville. 2016. *Deep Learning*. MIT Press.

Treatments, Outcomes, and Confounding: Core Concepts

1 Introduction

Causal inference [1] primarily concerns itself with understanding how one variable influences another. Before delving deeper into complex causal models, it is critical to precisely define the building blocks of causal reasoning: treatments, outcomes, and confounding. Treatments represent interventions or actions, outcomes are the effects we observe, and confounders (variables that influence both the treatment and the outcome, potentially biasing naive comparisons [2]). Confounders are variables that can obscure authentic causal relationships. Without clarity on these core concepts, attempts at causal analysis risk being misleading. This chapter develops these foundations systematically, combining intuition, formalism, and illustrative examples.

2 Treatments

Treatments refer to the variables we intervene upon or manipulate. In experimental sciences, treatments correspond to controlled factors, such as administering a drug or modifying a feature in a system. In observational settings, treatments are still identifiable but not directly controlled.

Formally, let T denote the treatment variable. Treatments can be either binary or continuous. A binary treatment is one where the intervention has two levels, typically denoted $T = 1$ (treated) and $T = 0$ (untreated). An example of a binary treatment is whether or not a patient receives a new medication. Continuous treatments, on the other hand, span a range of values. For instance, the dosage level of a drug can vary continuously rather than being a simple yes/no.

© The Author(s), under exclusive license to Springer Nature Switzerland AG 2025

D. Rajamanickam, *Causal Inference for Machine Learning Engineers*,

https://doi.org/10.1007/978-3-031-99680-1_2

Understanding the nature of the treatment is crucial because it shapes the estimation strategies later employed. For example, estimating causal effects for binary treatments often simplifies comparing two groups, whereas techniques such as dose-response curve modeling become necessary for continuous treatments.

Causal and ML

Causal Concepts for Machine Learning

Causal inference provides tools to move beyond correlational insights from machine learning to actionable knowledge for interventions.

- **Why Care About Causes?:** Machine learning is great at spotting patterns, but causal inference is about understanding the **why**. It's the difference between a weather app that shows you it always rains after sunrise, and one that explains **how** clouds cause rain. The second app is more useful.
- **Feature Importance: The Causal Lens:** Machine learning feature importance tells you which inputs **correlate** with predictions. Causal inference helps you find the features that **drive** the predictions. It's like finding the puppet master, not just the puppets.
- **The Enemy: Adversarial Attacks:** Adversarial examples in machine learning are like illusions. They fool the model because they exploit non-causal patterns. Causal models are harder to trick because they focus on real relationships.

---

3 Outcomes

Outcomes capture the effects resulting from the application of a treatment. Denote Y as the outcome variable of interest. The fundamental goal of causal inference is understanding how Y changes as a function of T.

Outcomes can vary in type. They may be continuous, such as blood pressure levels, or categorical, such as whether a patient experiences remission. Crucially, for each unit or individual in the study, we imagine two Potential Outcomes: $Y(1)$, the outcome if the unit receives the treatment, and $Y(0)$, the outcome if the unit does not receive the treatment. This is the basis of the *potential outcomes framework*, also called the Rubin Causal Model [3].

However, in practice, we can only observe one of these outcomes for each unit, corresponding to the treatment received. The unobserved outcome is the *counterfactual* [3]. This basic asymmetry underpins the challenges of causal inference.

Causal and ML

Treatments in the ML World

Understanding the treatment variable is essential for designing practical experiments and interventions, a process that can be informed by feature selection in machine learning.

- **Tuning Knobs:** Think of "treatments" as the "tuning knobs" of a machine learning model. What happens if we change the learning rate? What if we use a different activation function? Causal inference helps us understand the effect of these changes.
- **Feature Engineering Choices:** Feature engineering is like deciding what ingredients to give to the model. Causal inference helps us choose features that we can actively **control** to influence the outcome.
- **Reinforcement Learning Actions:** In reinforcement learning, "treatments" are an agent's actions. Causal inference is crucial for building agents that learn the **consequences** of their actions.

---

4 Confounding and Spurious Associations

Confounding arises when an external variable influences the treatment and the outcome, creating a spurious association that can be mistaken for causation. Formally, let Z denote a confounder. If Z affects both T and Y, then naive comparisons [2] between treated and untreated groups may reflect differences in Z rather than an actual causal effect of T on Y.

Causal and ML

Confounding in Machine Learning

Confounding variables are a major source of bias in machine learning datasets, and causal inference techniques offer strategies to mitigate their impact.

- **The Illusion of Cause:** Confounding is like a magician's trick. It makes it **seem** like the treatment causes the outcome, when it's something else. This is how ML models get fooled by spurious correlations.
- **The Sneaky Bias:** Confounding is like a hidden bias in the training data. The model learns this bias, and it affects its predictions.
- **Finding the Real Signal:** Causal inference gives us tools to "see through" the confounding and find the **true** effect of the treatment. It's like finding the real signal in a noisy channel.

A classical example involves smoking (T) and lung cancer (Y), with genetic predisposition (Z) acting as a confounder. Individuals genetically predisposed to nicotine addiction may both smoke more and have higher baseline risks of lung cancer, independently of smoking behavior. Failing to account for Z can lead to biased estimates of the causal effect of smoking.

Causal and ML

Outcomes and Model Behavior

Causal inference helps to define appropriate outcome variables that truly reflect the effect of a treatment, rather than being influenced by other factors.

- **Model's Voice:** "Outcomes" are like the model's "voice." What does it predict? What decisions does it make? Causal inference helps us understand how the "treatment" changes the model's behavior.
- **The "What If" Game:** Potential outcomes are like playing "What if?" with the model. What **would** it have predicted if we'd changed the input? This is counterfactual prediction.
- **Beyond Accuracy:** A high-accuracy model isn't always good. It might make bad decisions if it learned the wrong reasons (non-causal). Causal inference pushes us beyond simple accuracy.

To visualize confounding, consider the following directed acyclic graph (DAG).

Fig. 1 Causal diagram
illustrating confounding

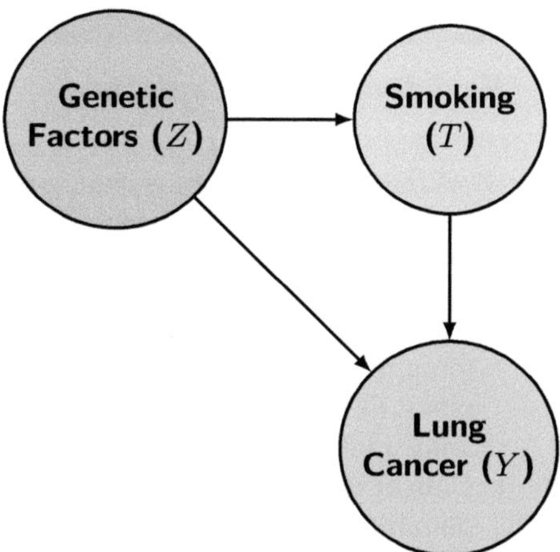

Here, Z acts as a common cause of both T and Y, forming a structure known as a *fork*. Adjusting for Z typically through regression, stratification, or matching is necessary to block the spurious path from T to Y that flows through Z (Fig. 1).

5 Backdoor Paths and Bias

In graphical terms, a *backdoor path* is any path from T to Y that starts with an arrow into T. Backdoor paths represent sources of spurious association due to confounding.

The *backdoor criterion*, introduced by Pearl [1], provides a formal condition under which a set of variables Z suffices to block all spurious paths:

Theorem 2.1 (Backdoor Criterion) *A set of variables Z satisfies the backdoor criterion [1] relative to (T, Y) if:*

1. *No node in Z is a descendant of T.*
2. *Z blocks every path between T and Y that contains an arrow into T.*

By conditioning on Z, we can recover the causal effect of T on Y even in the presence of confounding.

Practically, blocking a backdoor path removes the bias introduced by confounding, enabling valid causal inference. Failure to block backdoor paths leads to biased estimates and incorrect conclusions about cause and effect.

Causal and ML

Backdoor Paths Explained

Blocking backdoor paths is analogous to feature selection in machine learning, where we aim to identify the minimal set of features that predict the outcome without introducing bias.

- **The Leaky Faucet:** Imagine a faucet leaking water. The water flows through different pipes to the drain. Backdoor paths are like those extra pipes; they let "information" flow from the outcome to the treatment, creating a false connection.
- **Causal Graphs as Blueprints:** Causal graphs are like blueprints of the plumbing system. They help us see the backdoor paths and determine how to "block" them.
- **Regularization as a Filter:** Regularization in ML is like putting a filter on some pipes. It reduces the "leaky" information flow and helps the model focus on the main connection.

---

6   The Fundamental Problem of Causal Inference

At the heart of causal inference lies a profound challenge: for each unit, we can only observe the outcome under one treatment condition, not both. This is known as the *fundamental problem of causal inference*.

Formally, for a given individual i, we define:

$$Y_i(1) \quad \text{(potential outcome under treatment)}$$

$$Y_i(0) \quad \text{(potential outcome under no treatment)}$$

But only one of these is observed, depending on whether $T_i = 1$ or $T_i = 0$. The missing potential outcome is counterfactual.

This missing data structure implies that causal effects cannot be directly measured at the individual level. Instead, we aim to estimate average causal effects across populations, such as the *Average Treatment Effect* (ATE):

$$\text{ATE} = \mathbb{E}[Y(1) - Y(0)]$$

Various identification strategies, such as randomization, stratification, instrumental variables [4], and propensity score [5] matching, are developed to overcome this fundamental limitation.

> **Causal and ML**
>
> ### *The Missing Piece*
>
> Estimating average treatment effects is a common goal in machine learning applications like A/B testing, where we want to measure the impact of changes on user behavior.
>
> - **The Parallel Universe:** The fundamental problem is like being unable to see into a parallel universe. We only see what **did** happen, not what *could have* happened.
> - **Estimating the Unknown:** Causal inference is about carefully **estimating** what would have happened in that parallel universe (the counterfactual). It's like a detective reconstructing a crime scene.
> - **Average Effects:** Because we can't know the exact effect on each individual, we often focus on the **average** effect across a group. This is like estimating the average height of people in a city.

7 Examples of Confounding in Real Life

Example 1: Education and Income

Suppose we wish to study the causal effect of education (T) on income (Y). Individuals who pursue more education may differ systematically from those who do not, for example, regarding family background or innate ability (Z). These factors influence both the likelihood of acquiring education and future earnings. Failing to adjust for such confounders could lead to overstating or understating the true causal effect of education.

Example 2: Exercise and Health

Consider the relationship between exercise (T) and overall health outcomes (Y). Health consciousness (Z) acts as a confounder: more health-conscious individuals are more likely to exercise and adopt other healthy behaviors such as eating nutritious food. Observed correlations between exercise and health could thus partially reflect differences in underlying health attitudes rather than the effect of exercise alone.

These examples underline the importance of identifying, measuring, and adjusting for confounders when attempting to make causal claims.

Causal and ML

Confounding in ML Scenarios

Real-world machine learning datasets are rife with confounding, emphasizing the need for causal reasoning to build reliable models.

- **Recommendations and Popularity:** A recommender system might show that showing a movie (treatment) leads to higher ratings (outcome). But maybe popular movies (confounder) get shown **and** get high ratings.
- **Language and Sentiment:** A sentiment analysis model might think that using certain words (treatment) **causes** positive sentiment (outcome). But maybe the topic of the text (confounder) influences both the words and the sentiment.
- **The Need for Context:** Confounding shows why machine learning models need **context**. They can't just look at simple correlations; they must understand the bigger picture.

---

8   Chapter Summary

Treatments, outcomes, and confounders [2].

From the basic vocabulary of causal inference, treatments represent variables we manipulate or study, outcomes capture the effects of these manipulations, and confounders introduce bias by linking treatments and outcomes through backdoor paths. Recognizing the fundamental problem of causal inference—the missing counterfactual—motivates the need for careful design and analysis strategies to estimate causal effects. With these foundational concepts in place, the following chapters will develop formal methods for causal estimation.

9   Code

Listing 2.1   Chapter 2: Focusing on simulating confounding and demonstrating its effect

```python
import numpy as np
import pandas as pd
import matplotlib.pyplot as plt

# --- Basic Simulation ---
np.random.seed(1)  # For reproducibility
n = 1000  # Number of data points
```

```python
# Confounder: Health Consciousness (higher values = more
    health-conscious)
health_consciousness = np.random.normal(0, 1, n)

# Treatment: Aspirin Use (1 = uses aspirin, 0 = doesn't)
# Health-conscious people are more likely to use aspirin
aspirin_probability = 1 / (1 + np.exp(-0.5 *
    health_consciousness))  # Sigmoid function to get
    probability
aspirin = np.random.binomial(1, aspirin_probability, n)

# Outcome: Headache Relief (1 = relief, 0 = no relief)
# Health consciousness also independently affects headache
    relief
relief_probability = 0.2 + 0.4 * aspirin + 0.3 *
    health_consciousness + np.random.normal(0, 0.2, n)
relief = np.random.binomial(1, np.clip(relief_probability,
    0, 1), n)  # Clip to ensure probabilities are valid

# Create Pandas DataFrame
data = pd.DataFrame({'HealthConsciousness':
    health_consciousness,
                     'Aspirin': aspirin,
                     'Relief': relief})

# Calculate and print the observed association (not causal!)
observed_association = data.groupby('Aspirin')['Relief'].
    mean()
print("\nObserved Association (Aspirin vs. Relief):")
print(observed_association)

# Visualize the data (example: histograms)
plt.figure(figsize=(10, 5))

plt.subplot(1, 2, 1)
plt.hist(data[data['Aspirin'] == 0]['Relief'], alpha=0.5,
    label='No Aspirin')
plt.hist(data[data['Aspirin'] == 1]['Relief'], alpha=0.5,
    label='Aspirin')
plt.xlabel('Relief (0 or 1)')
plt.ylabel('Frequency')
plt.title('Relief Distribution by Aspirin Use')
plt.legend()

plt.subplot(1, 2, 2)
plt.scatter(data['HealthConsciousness'], data['Aspirin'],
    alpha=0.5)
plt.xlabel('Health Consciousness')
plt.ylabel('Aspirin Use')
plt.title('Aspirin Use vs. Health Consciousness')

plt.show()
```

```python
print("\n--- Explanation ---")
print("This code simulates the effect of a confounding
    variable ('HealthConsciousness') on both the treatment
    ('Aspirin') and the outcome ('Relief').")
print("1.  'HealthConsciousness' is generated from a normal
    distribution.")
print("2.  'Aspirin' use is made *dependent* on '
    HealthConsciousness' (health-conscious people are more
    likely to take it).")
print("3.  'Relief' is influenced by *both* 'Aspirin' and '
    HealthConsciousness'.")
print("The plots and the 'observed_association' show that
    aspirin appears to be associated with relief, but this
    association is biased because of 'HealthConsciousness'."
    )
print("In a real-world scenario, if we don't account for '
    HealthConsciousness', we might wrongly conclude that
    aspirin is more effective than it actually is.")


# --- Structured Simulation with Functions (More Advanced)
    ---
def simulate_confounded_data(n=1000):
    """Simulates data with a confounding variable."""

    health = np.random.normal(0, 1, n)
    aspirin_prob = 1 / (1 + np.exp(-0.5 * health))
    aspirin = np.random.binomial(1, aspirin_prob, n)
    relief_prob = 0.2 + 0.4 * aspirin + 0.3 * health + np.
        random.normal(0, 0.2, n)
    relief = np.random.binomial(1, np.clip(relief_prob, 0,
        1), n)

    return pd.DataFrame({'Health': health, 'Aspirin':
        aspirin, 'Relief': relief})


def analyze_association(data):
    """Calculates and prints the association between aspirin
        use and relief."""

    association = data.groupby('Aspirin')['Relief'].mean()
    print("\n--- Analysis: Aspirin vs. Relief ---")
    print(association)


def visualize_confounding(data):
    """Visualizes the relationship between health
        consciousness and aspirin use."""

    plt.figure(figsize=(6, 6))
    plt.scatter(data['Health'], data['Aspirin'], alpha=0.5)
    plt.xlabel('Health Consciousness')
```

```python
    plt.ylabel('Aspirin Use')
    plt.title('Confounding: Health vs. Aspirin')
    plt.show()


if __name__=="__main__":  # Ensures code only runs when the script
    is executed directly
    data = simulate_confounded_data()
    analyze_association(data)
    visualize_confounding(data)

    print("\n--- Function-Based Explanation ---")
    print("The code is organized into functions for better
        readability and reusability.")
    print("1.  'simulate_confounded_data()' generates the
        data, encapsulating the simulation logic.")
    print("2.  'analyze_association()' calculates and prints
        the mean relief for aspirin users and non-users.")
    print("3.  'visualize_confounding()' creates a scatter
        plot to show the relationship between 'Health' and '
        Aspirin'.")
    print("The 'if__name__=="__main__":' block ensures that
        these functions are called only when the script is
        run, not when imported as a module.")
```

10 Exercises

1. Define treatment, outcome, and confounder in the context of causal inference.
2. Provide an example of a confounding variable and explain how it affects causal conclusions.
3. Why is adjusting for confounders critical when estimating treatment effects?
4. What happens if we adjust for a collider variable instead of a confounder?

References

1. Pearl, Judea. 2009. *Causality: Models, Reasoning, and Inference*, 2nd ed. Cambridge University Press.
2. Imbens, Guido W., and Donald B. Rubin. 2015. *Causal Inference for Statistics, Social, and Biomedical Sciences*. Cambridge University Press.
3. Rubin, Donald B. 1974. Estimating causal effects of treatments in randomized and nonrandomized studies. *Journal of Educational Psychology* 66 (5): 688–701.
4. Angrist, Joshua D, Guido W Imbens, and Donald B Rubin. 1996. Identification of causal effects using instrumental variables. *Journal of the American Statistical Association* 91 (434): 444–455.
5. Rosenbaum, Paul R, and Donald B Rubin. 1983. The central role of the propensity score in observational studies for causal effects. *Biometrika* 70 (1): 41–55.

Causal Estimation Basics

1 Introduction

Causal inference [1] is not merely about describing associations or correlations we observe in data. Instead, it is primarily concerned with understanding what happens when we intervene in the world: how outcomes change if we manipulate some variable. This manipulation aspect is what distinguishes causality from prediction.

Imagine a healthcare researcher who wants to know whether a new drug reduces blood pressure. Simply observing that patients who take the drug have lower blood pressure does not answer the causal question, because these patients might differ in other ways. Perhaps they exercise more or eat healthier diets. Therefore, we must carefully separate correlation from true causal effects.

In this chapter, we lay the foundation for causal estimation. We introduce key concepts such as potential outcomes and the average treatment effect and formalize the assumptions needed to make valid causal claims from data. We also explain the limitations of observational studies and how statistical methods attempt to recover causal relationships despite those limitations.

Causal estimation provides a framework for building machine learning models that predict well and generalize to new situations where the treatment is applied.

D. Rajamanickam, *Causal Inference for Machine Learning Engineers*,
https://doi.org/10.1007/978-3-031-99680-1_3

> **Causal and ML**
>
> ### *Causal Estimation for ML Engineers*
>
> - **Beyond Correlations:** Machine learning often stops at finding correlations: "These features predict that outcome." Causal inference pushes further: "Changing these features will **cause** this change in the outcome."
> - **The Intervention Focus:** Causal estimation is like having a control knob on the world. We want to know: "If I turn this knob (change the input), how much will the dial (output) move?"
> - **Why This Matters for ML:** If our machine learning model is going to **make decisions** or **recommend actions**, we need causal estimates. Otherwise, we're just guessing about the consequences.

2 Problem Setup

Suppose we are interested in understanding the effect of a treatment, which we denote by T, on an outcome, denoted by Y. Treatments can be binary (e.g., did you take the drug or not) or continuous (e.g., dosage level). The goal is to measure how changing T causes a change in Y.

To reason carefully about causality, we introduce the concept of **potential outcomes** for each individual:

- $Y(1)$: the outcome if the individual receives the treatment.
- $Y(0)$: the outcome if the individual does not receive the treatment.

However, we can only observe one of these outcomes for any individual. If a patient takes the drug, we observe $Y(1)$; we cannot simultaneously observe what would have happened if they had not taken the drug.

This dilemma is known as the **fundamental problem of causal inference**. We can never directly observe causal effects at the individual level; we can only reason about them statistically across groups.

Thus, the causal effect for an individual is:

$$\tau_i = Y_i(1) - Y_i(0)$$

But since we cannot observe both $Y_i(1)$ and $Y_i(0)$ for the same individual, our goal shifts to estimating **average effects** across a population.

Causal and ML

Potential Outcomes: The ML Analogy

Potential outcomes are closely related to counterfactual predictions in machine learning, where we want to estimate what the model would predict under different inputs.

- **The Model's Two Faces:** Think of $Y(1)$ and $Y(0)$ as two different predictions our machine learning model **could** make for the same input. $Y(1)$ is the prediction if we **force** the treatment feature to be "on," and $Y(0)$ is the prediction if we force it to be "off."
- **The Unseen Prediction:** The fundamental problem is that we only see one of these predictions in the real world. The other one is the "what if" prediction, the counterfactual.
- **Personalized ML:** Estimating the difference between $Y_i(1)$ and $Y_i(0)$ is like personalizing a machine learning model for each user or situation. Some users might respond strongly to a treatment (e.g., an ad), while others might not.

Observed Data

For each individual i, we observe:

$$(X_i, T_i, Y_i)$$

where:

- X_i are pre-treatment characteristics or covariates (such as age, gender, income),
- T_i is the treatment assignment (0 or 1),
- Y_i is the observed outcome.

But we cannot observe both potential outcomes for the same individual—only $Y_i(T_i)$, depending on whether $T_i = 0$ or 1.

3 What We Observe and What We Want

When we look at data, it is tempting to compare the average outcome among treated individuals to that among untreated individuals:

$$\mathbb{E}[Y \mid T = 1] - \mathbb{E}[Y \mid T = 0]$$

However, this comparison is not necessarily causal. Treated and untreated individuals may differ systematically in ways that affect Y. For example, sicker patients may be more likely to seek treatment.

What we truly want is the **Average Treatment Effect (ATE)** [2], defined as:

$$\text{ATE} = \mathbb{E}[Y(1) - Y(0)]$$

This quantity captures the average change in outcomes if, hypothetically, we treated everyone versus treated no one.

Example: New Training Program

Imagine a company introduces a new employee training program and wants to know its effect on productivity. Employees who choose to participate may be more motivated than others. Thus, comparing average productivity between trained and untrained employees would confound the impact of training with the employees' inherent motivation.

To estimate the causal effect of the training program, we must account for such differences.

Causal and ML

From Simple Averages to Causal Truth

Estimating the ATE is similar to evaluating the average performance of a machine learning model across different user groups or experimental conditions.

- **The Naive Model:** Comparing $\mathbb{E}[Y|T = 1]$ and $\mathbb{E}[Y|T = 0]$ is like building the simplest possible machine learning model: just averaging the outcomes for treated and untreated groups. It's often a bad model because it ignores other factors.
- **The Ideal Target:** The ATE ($\mathbb{E}[Y(1) - Y(0)]$) is the **ideal** thing we want our model to estimate. It's the average effect of the treatment across everyone, if we could somehow see both treated and untreated outcomes for everyone.
- **The Real-World Mess:** The challenge is that the real world is messy. Treated and untreated groups are usually different in other ways, which confuses our simple average model.

4 Estimating Average Treatment Effects

Because we can never observe both potential outcomes for the same individual, causal estimation from observational data relies on assumptions.

Two key assumptions are essential.

4.1 Unconfoundedness (Ignorability)

Formally:

$$(Y(0), Y(1)) \perp\!\!\!\perp T \mid X$$

This means that after conditioning on observed covariates X, the assignment to treatment T is as good as random. In other words, individuals with the same X have the same probability of receiving treatment, independent of their potential outcomes.

Example: If we know a patient's age, gender, and pre-existing conditions, treatment assignment is independent of their unobserved health outcomes.

Causal and ML

Ignorability: The "As Good As Random" Idea

Violations of ignorability can lead to biased machine learning models, just as they can bias causal effect estimates.

- **The Perfect Experiment:** Ignorability is like saying that, *within each group of people with the same X*, the treatment was assigned randomly. It's like having lots of little perfect experiments.
- **Feature Completeness:** For ignorability to be plausible, our features X need to capture **everything** that's relevant to treatment assignment. It's like giving our machine learning model **all** the important input information.
- **The Real-World Gap:** In reality, ignorability is often an assumption we **hope** is roughly true, not something we can guarantee.

4.2 Positivity

Formally:

$$0 < P(T = 1 \mid X) < 1$$

for all X.

This ensures that every type of individual (defined by X) has a nonzero probability of being treated or untreated. Without positivity, causal effects are not identifiable in those regions of the covariate space.

Violation Example

If only high-risk patients ever receive the drug (and low-risk patients never do), then we cannot estimate the drug's effect on low-risk patients—positivity is violated.

Causal and ML

Positivity: Enough Data for Everyone

The positivity assumption is crucial for ensuring that machine learning models have sufficient data to learn the treatment effect across the entire range of covariates

- **Data Representation:** Positivity is like making sure our training data has examples of **everyone** getting **every** treatment. No group is completely excluded.
- **Model Extrapolation:** If we violate positivity, our model is forced to **guess** what would happen for people it's never seen treated (extrapolation), which is dangerous in machine learning.
- **The Real-World Constraint:** In practice, positivity might not hold perfectly, but we try to get as close as possible.

---

5 Strategies for Causal Estimation

Assuming ignorability and positivity hold, we have several strategies to estimate causal effects.

5.1 *Regression Adjustment*

We model the outcome Y as a function of covariates and treatment:

$$Y = f(X, T) + \epsilon$$

For example, using linear regression:

$$Y = \beta_0 + \beta_1 T + \beta_2 X + \epsilon$$

Here, β_1 represents the causal effect of T on Y, adjusting for X.

Causal and ML

Regression: A Simple Causal Model

Regression adjustment can be seen as a simple form of causal modeling, where we use regression coefficients to estimate the effect of the treatment on the outcome.

- **Linear Cause-and-Effect:** Linear regression is like assuming a simple, direct cause-and-effect relationship. The coefficient for the treatment is our estimate of the treatment's effect.
- **Model Limitations:** If the real relationship is curved or complex, linear regression will give a distorted (biased) answer. This is like using a blurry lens in a camera.
- **ML Extension:** We can use more flexible machine learning models (e.g., neural networks) for the regression, but we must still be careful about causal interpretation.

5.2 Propensity Score Methods

The propensity score [2] is defined as:

$$e(X) = P(T = 1 \mid X)$$

It summarizes the probability of receiving treatment given covariates. Propensity scores can be used to:

- Match treated and untreated individuals with similar $e(X)$,
- Stratify data into bins of similar $e(X)$,
- Weight observations by the inverse of $e(X)$ (IPW).

> **Important**
>
> **Propensity Scores: Balancing the Groups**
>
> Machine learning models can estimate propensity scores, providing a powerful tool for balancing treated and untreated groups in observational studies.
>
> - **Matching People:** Propensity scores are like a way to "match" treated and untreated people who are very similar. It's like creating balanced teams for a game.
> - **Re-weighting the Data:** We can also use propensity scores to re-weight the data, giving more importance to people who are unusual in terms of who got treated.
> - **ML's Role:** Machine learning models can be used to **predict** these propensity scores, helping us balance the data.

5.3 *Inverse Probability Weighting (IPW)*

Using weights derived from the propensity score, we can create a pseudo-population where treatment is independent of covariates.

The IPW estimator for ATE (The potential outcomes framework), introduced by Rubin [3], formalizes the concept of individual treatment effects and average treatment effects [3].

$$\text{ATE} = \mathbb{E}\left[\frac{T \cdot Y}{e(X)} - \frac{(1 - T) \cdot Y}{1 - e(X)}\right]$$

This reweights observations so that treated and untreated groups are comparable.

> **Causal and ML**
>
> **IPW: Correcting for Biased Sampling**
>
> IPW is a re-weighting technique that can be used to correct for biases in machine learning datasets, similar to how it's used in causal inference.
>
> - **Sampling Bias:** IPW is like correcting for a biased way of sampling data. If some people were much more likely to be included in our study, IPW gives them less weight.
> - **Model Instability:** IPW can sometimes make our estimates sensitive to a few data points, especially if some propensity scores are close to 0 or 1. This is like having a machine learning model that's very unstable.
> - **Regularization Analogy:** We often use techniques similar to regularization in machine learning to make IPW more stable.

6 The Role of Covariates

Covariates X are crucial in causal inference because they allow us to control for confounding variables [2] that would otherwise bias our estimates.

However, adjusting for the wrong covariates (e.g., colliders or post-treatment variables) can introduce new biases. Causal graphs (DAGs) help visualize and reason which covariates to adjust for—a topic we explore more deeply in the next chapter.

> **Causal and ML**
>
> ### *Covariates: The Context for Causality*
>
> Careful selection of covariates is essential for both causal inference and machine learning; including irrelevant covariates can add noise, while omitting key covariates can introduce bias.
>
> - **Causal Context:** Covariates X give us the context for understanding the treatment's effect. It's like saying, "The treatment has this effect *for this type of person*."
> - **Feature Selection:** Choosing the right covariates is like choosing the right features for a machine learning model. If we include irrelevant features, we add noise. If we exclude important features, we get a biased answer.
> - **Causal Graphs as Guides:** Causal graphs (from the next chapter) can help us decide which covariates to include, like a map showing us the right path.

7 A Complete Example

Suppose we simulate data as follows:

$$X \sim \mathcal{N}(0, 1) \quad \text{(covariates)}$$
$$T \sim \text{Bernoulli}(\sigma(0.5X)) \quad \text{(treatment)}$$
$$Y(0) = X + \epsilon_0 \quad \epsilon_0 \sim \mathcal{N}(0, 1)$$
$$Y(1) = X + 2 + \epsilon_1 \quad \epsilon_1 \sim \mathcal{N}(0, 1)$$

Thus, the actual treatment effect is constant and equal to 2.

However, since T depends on X, comparing treated and untreated outcomes will be biased. We can fit a model adjusting for X, or use propensity scores, to recover the true effect (Fig. 1).

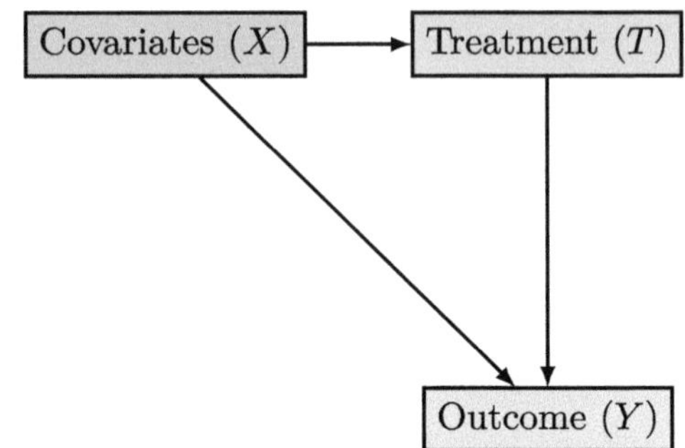

Fig. 1 Simple causal DAG showing confounding between treatment and outcome through covariates

> **Causal and ML**
>
> ### *Simulation: A Causal Sandbox*
>
> Simulating data with known causal effects provides a valuable way to validate machine learning models designed for causal inference.
>
> - **Testing Our Tools:** The simulation is like a sandbox where we can test our causal inference tools. We know the "true" answer, so we can see if our methods work.
> - **ML Validation:** This is similar to how we use simulated data to test machine learning algorithms, but here, we're validating that we're getting the **causal** effect right.

8   Chapter Summary

In this chapter, we built the basic framework for causal estimation. We introduced potential outcomes, average treatment effects, , and the key assumptions required for identifying causal effects. We discussed different methods, including regression adjustment, propensity scores [4], and inverse probability weighting, and saw how covariates X play a central role in mitigating biases.

The next chapter will introduce causal graphs (Directed Acyclic Graphs, or DAGs) as a powerful tool for explicitly representing causal assumptions.

9 Code

Listing 3.1 Chapter 3: Focusing on estimating the Average Treatment Effect (ATE)

```python
import numpy as np
import numpy as np
import pandas as pd
import statsmodels.formula.api as sm  # For regression
from sklearn.linear_model import LogisticRegression  # For
    propensity score
from sklearn.utils import resample  # For bootstrapping

import matplotlib.pyplot as plt
import seaborn as sns  # For better plots


# --- 1. Simulate Data with Confounding ---
def simulate_confounded_data(n=1000, ate=5):
    """Simulates data with a treatment, outcome, and
        confounder."""

    X = np.random.normal(0, 1, n)  # Confounder
    T = np.random.binomial(1, 1 / (1 + np.exp(-0.5 * X)), n)
        # Treatment depends on X
    Y0 = 2 * X + np.random.normal(0, 1, n)  # Outcome
        without treatment
    Y1 = Y0 + ate  # Outcome with treatment (constant ATE)
    Y = T * Y1 + (1 - T) * Y0  # Observed outcome

    return pd.DataFrame({'X': X, 'T': T, 'Y': Y, 'Y0': Y0, '
        Y1': Y1})  # Return potential outcomes too


# --- 2. Estimate ATE using Regression Adjustment ---
def estimate_ate_regression(data):
    """Estimates ATE using linear regression."""

    model = sm.ols('Y ~ T + X', data=data).fit()  # Adjust
        for X
    ate = model.params['T']  # Coefficient of T is the ATE
    print("\n--- Regression Adjustment ---")
    print(f"Estimated ATE: {ate:.2f}")

    return ate, model


# --- 3. Estimate ATE using Propensity Score Weighting (IPW)
    ---
def estimate_ate_ipw(data, bootstrap_iterations=1000):
    """Estimates ATE using Inverse Propensity Weighting (IPW
        )."""

    # 3.1. Estimate propensity scores
```

```python
    propensity_model = LogisticRegression(solver='liblinear'
        , random_state=0).fit(data[['X']], data['T'])
    propensity_scores = propensity_model.predict_proba(data
        [['X']])[:, 1]
    data['Propensity'] = propensity_scores

    # 3.2. Calculate IPW weights
    data['IPW'] = np.where(data['T'] == 1, 1 / data['
        Propensity'], 1 / (1 - data['Propensity']))

    # 3.3. Estimate ATE with weighted average
    ate_ipw = np.average(data['Y'], weights=data['IPW']) -
        np.average(data['Y'], weights=np.where(data['T'] ==
        0, data['IPW'], 0))
    print("\n--- IPW ---")
    print(f"Estimated ATE (IPW): {ate_ipw:.2f}")

    # 3.4. (Optional) Bootstrap for Confidence Intervals
    ate_estimates = []
    for _ in range(bootstrap_iterations):
        bootstrap_data = resample(data, replace=True,
            n_samples=len(data), random_state=np.random.
            randint(0, 10000))  # Resample with replacement
        ate_estimates.append(np.average(bootstrap_data['Y'],
            weights=bootstrap_data['IPW']) - np.average(
            bootstrap_data['Y'], weights=np.where(
            bootstrap_data['T'] == 0, bootstrap_data['IPW'],
            0)))

    ci_lower = np.percentile(ate_estimates, 2.5)
    ci_upper = np.percentile(ate_estimates, 97.5)
    print(f"95% Confidence Interval: ({ci_lower:.2f}, {
        ci_upper:.2f})")  # More robust uncertainty

    return ate_ipw, propensity_model


# --- 4. Visualize Results ---
def visualize_results(data, regression_model,
    propensity_model):
    """Visualizes the data and the propensity score
        distribution."""

    plt.figure(figsize=(15, 5))

    # 4.1. Scatter Plot: Y vs. T, colored by X
    plt.subplot(1, 3, 1)
    plt.scatter(data['T'], data['Y'], c=data['X'], alpha
        =0.5)
    plt.xlabel('Treatment (T)')
    plt.ylabel('Outcome (Y)')
    plt.title('Y vs. T (colored by X)')
    plt.colorbar(label='Confounder (X)')
```

```python
    # 4.2. Regression Line
    plt.subplot(1, 3, 2)
    sns.regplot(x='T', y='Y', data=data, ci=95)  # Show
        regression line with CI
    plt.xlabel('Treatment (T)')
    plt.ylabel('Outcome (Y)')
    plt.title('Regression of Y on T')

    # 4.3. Propensity Score Distribution
    plt.subplot(1, 3, 3)
    plt.hist(data['Propensity'], bins=20, alpha=0.5)
    plt.xlabel('Propensity Score')
    plt.ylabel('Frequency')
    plt.title('Propensity Score Distribution')

    plt.tight_layout()  # Adjust layout to prevent
        overlapping
    plt.show()


# --- 5. Main Execution ---
if __name__=="__main__":
    np.random.seed(42)  # Consistent seed for the whole
        script
    data = simulate_confounded_data()
    ate_regression, regression_model = 
        estimate_ate_regression(data)
    ate_ipw, propensity_model = estimate_ate_ipw(data)
    visualize_results(data, regression_model,
        propensity_model)

    print("\n--- Summary ---")
    print("This code simulates data with a confounder 'X'
        affecting both treatment 'T' and outcome 'Y'.")
    print("It then estimates the Average Treatment Effect
        (ATE) using two methods:")
    print("1. Regression Adjustment: Includes 'X' as a
        covariate in the regression model.")
    print("2. Inverse Propensity Weighting (IPW): Weights
        observations by the inverse probability of treatment
        .")
    print("The visualization shows the confounding effect
        and the distribution of propensity scores.")
```

10 Exercises

1. Explain what an Average Treatment Effect (ATE) is.
2. Define Individual Treatment Effect (ITE) and how it differs from ATE.

3. Why can't we directly observe an individual's ITE in practice?
4. Give an example where estimating the ATE is insufficient for decision-making.

References

1. Pearl, Judea. 2009. *Causality: Models, Reasoning, and Inference*. 2nd ed. Cambridge University Press.
2. Imbens, Guido W., and Donald B. Rubin. 2015. *Causal Inference for Statistics, Social, and Biomedical Sciences*. Cambridge University Press.
3. Rubin, Donald B. 1974. Estimating causal effects of treatments in randomized and nonrandomized studies. *Journal of Educational Psychology* 66 (5): 688–701.
4. Rosenbaum, Paul R., and Donald B Rubin. 1983. The central role of the propensity score in observational studies for causal effects. *Biometrika* 70 (1): 41–55.

Causal Graphs: Structure and Assumptions

1 Introduction

Causal graphs offer a powerful and intuitive way to represent assumptions about cause-and-effect relationships among variables. They use directed edges to encode causal relationships and provide a visual framework for reasoning about the flow of causality, detecting biases, and designing strategies for causal estimation.

Rather than relying solely on equations, causal graphs allow researchers to communicate assumptions clearly, aiding both qualitative reasoning and quantitative modeling. In this chapter, we develop an understanding of causal graphs and learn how to construct and interpret them.

Causal and ML

Causal Graphs for Machine Learning

Causal graphs provide a structured way to represent the flow of information and dependencies within machine learning models.

- **Visualizing the Model's World:** Causal graphs are like roadmaps for a machine learning model. They show how the model **thinks** the world works, which features influence others, and how they ultimately affect the prediction.
- **Feature Interactions as Paths:** In machine learning, we often talk about feature interactions. Causal graphs clarify these interactions, showing how one feature's effect might flow through other features.
- **Debugging Bias:** If a machine learning model makes biased predictions, a causal graph can help us trace the source of the bias. Which features are causing it? How are they connected?

D. Rajamanickam, *Causal Inference for Machine Learning Engineers*,
https://doi.org/10.1007/978-3-031-99680-1_4

2 What Is a Causal Graph?

A **causal graph** is a *directed acyclic graph* (DAG) [1] that provides a graphical representation of causal assumptions [1]. Pearl [1] Where:

- **Nodes** represent variables in the system.
- **Directed edges** (arrows) represent direct causal relationships.

Directed means that edges have a direction, indicating the direction of causal influence. **Acyclic** means that there are no loops; starting from any node and following the direction of the arrows, one cannot return to the starting node.

> **Key Idea**
>
> A causal graph encodes our assumptions about which variables causally affect others. It does not emerge automatically from the data. Rather, it must be constructed based on domain knowledge and prior studies.

> **Causal and ML**
>
> ### DAGs: The Flow of Information
>
> - **Nodes as Model Inputs/Outputs:** In a machine learning context, nodes can represent the input features, the intermediate calculations within the model, or the final prediction.
> - **Edges as Model Connections:** Edges show how information flows through the model. For example, an edge from "age" to "income" means the model uses age to predict income.
> - **Acyclicity and Feedback Loops:** The "acyclic" part means no feedback loops exist. This is important for understanding how information is processed in the model.

3 Nodes: Variables in the System

Each node in a causal graph represents a variable of interest. These variables could be observed or unobserved. In machine learning, nodes can represent:

- *Treatment variables*: whether a patient receives a drug.
- *Outcome variables*: such as patient survival.
- *Confounders*: such as age, socioeconomic status, or genetic predisposition.

Circles or rectangles typically represent nodes. The goal is to include all relevant variables needed to accurately represent the system.

Causal and ML

Nodes in Machine Learning

- **Input Features:** Nodes often represent the raw data fed into a machine learning model (e.g., pixels in an image, words in a sentence).
- **Hidden Units:** In neural networks, nodes can also represent the "hidden units" or intermediate calculations that the network performs.
- **Predictions:** The final output node represents the model's prediction.

4 Edges: Causal Relationships

An arrow from node A to node B ($A \to B$) indicates that A has a direct causal effect on B.

Notably, the lack of an edge implies the absence of a direct causal effect, conditional on the other variables included in the graph. If two variables are connected indirectly through a chain of arrows, their relationship is mediated through intermediate variables.

5 Building Causal Graphs: An Example

Suppose we are studying the effect of exercise on weight loss. A simple causal graph may look like (Fig. 1).

However, real-world systems are rarely this simple. Diet and socioeconomic status (SES) may also play important roles (Fig. 2).

Fig. 1 Simple causal graph: exercise causally affects weight loss

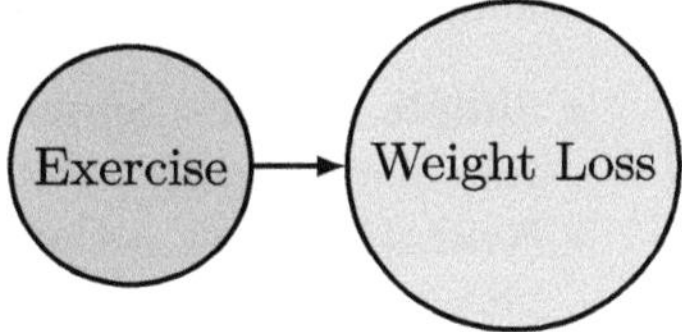

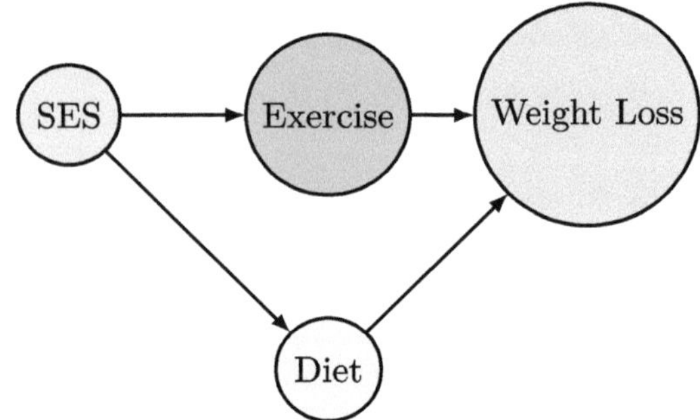

Fig. 2 More realistic causal graph with confounding and mediating paths

---

Causal and ML

Edges as Model Mechanisms

Edges illustrate how a machine learning model uses certain inputs to compute other values, ultimately leading to its prediction.

- **Model's "Reasoning":** Edges represent the model's assumed causal relationships. They show how the model "reasons" from inputs to outputs.
- **Direct Versus Indirect Effects:** An edge shows a **direct** effect. If there's no edge, any relationship is **indirect**, flowing through other nodes.
- **Absence of Edge is Important:** The **absence** of an edge is a strong claim: the model assumes that variable *does not* directly influence that other variable.

Causal and ML

Graphing a Machine Learning Model

Visualizing a machine learning model as a causal graph makes its underlying assumptions about feature relationships explicit.

- **Simple Model = Simple Graph:** A basic linear regression model has a very simple graph: input features all point directly to the output.
- **Neural Network = Complex Graph:** A neural network has a more complex graph, with layers of nodes and edges showing how information is transformed.
- **Graphing Assumptions:** Drawing the graph forces us to be explicit about the model's assumptions: What directly is used to make a prediction? What does it ignore?

6 Paths in Causal Graphs

A **path** is a sequence of nodes connected by edges, regardless of edge direction.

- **Causal paths** transmit causal effects from cause to effect.
- **Backdoor paths** [1] transmit spurious associations due to confounding.

Understanding paths helps us decide whether observed associations reflect true causality or are confounded.

Causal and ML

Paths and Information Flow

Understanding paths is crucial for debugging machine learning models and identifying potential sources of bias.

- **Causal Paths = Model's Chain of Thought:** A causal path is like the model's "chain of thought," showing how information is passed from one calculation to the next.
- **Backdoor Paths = Model's Blind Spots:** Backdoor paths are like hidden shortcuts the model takes, leading to biased predictions. They're like the model's "blind spots."
- **Understanding Model Behavior:** By analyzing paths, we can better understand **why** a model makes certain predictions.

---

7 Common Structures in Causal Graphs

Three key structures frequently appear in causal graphs.

7.1 *Chain Structure (Mediation)*

A chain occurs when:
$$A \to B \to C$$

Here, B mediates the effect of A on C. Conditioning on B blocks the transmission of the effect from A to C.

> **Causal and ML**
>
> **Chains and Intermediate Layers**
>
> - **Neural Network Layers:** In a neural network, chain structures are like the layers of the network. Each layer processes the information from the previous layer.
> - **Mediation in Models:** Mediation shows how one feature's effect is **mediated** by another. For example, "input image" −> "edge detection layer" −> "object recognition layer" −> "final classification."

7.2   Fork Structure (Confounding)

A fork occurs when:

$$A \leftarrow B \rightarrow C$$

Here, B is a common cause (confounder) of both A and C. Failing to control for B may induce a spurious association between A and C.

> **Causal and ML**
>
> **Forks and Shared Inputs**
>
> - **Shared Input Branches:** Forks happen when a model uses the same input for multiple calculations. For example, "input text" −>"sentiment analysis branch" and "topic classification branch."
> - **Confounding in Representations:** Forks can create confounding **within** the model's representation.

7.3   Collider Structure

A collision occurs when:

$$A \rightarrow B \leftarrow C$$

Here, B is a variable caused by both A and C. Conditioning on B opens a path between A and C, introducing spurious association even if A and C are otherwise independent.

> **Important**
>
> Do not condition on colliders unless explicitly necessary. Doing so can introduce bias.

> **Causal and ML**
>
> **Colliders and Information Convergence**
>
> - **Merging Information:** Colliders happen when information from different parts of the model is merged. For example, "text features" $->$ "decision layer" $<-$ "image features."
> - **Collider Bias in Models:** Collider structures can introduce unexpected correlations in the model's calculations.

These structures have direct analogies in machine learning architectures and can help us understand how information is processed.

8 Rules of Reading Causal Graphs

The most crucial concept is **d-separation**, which determines whether two variables are independent given a set of conditioning variables.

- A path is **blocked** if it contains a collider that is not conditioned on.
- A path is also blocked if we condition on a mediator along the path.
- Otherwise, the path is **open** and can transmit association.

The rules of reading graphs help in deciding which variables to control for in order to eliminate bias.

> **Causal and ML**
>
> **_d-separation: Model Independence_**
>
> The rules of d-separation provide a formal way to analyze the conditional independence of variables within a machine learning model.
>
> - **Model's Conditional Independence:** d-separation helps us understand when different parts of the model are **independent** of each other, given certain conditions.
> - **Simplifying Model Analysis:** It's like finding parts of the model that operate separately, simplifying our analysis.
> - **Debugging Model Errors:** d-separation can help us pinpoint where errors originate in the model.

9 Why Use Causal Graphs?

Causal graphs provide:

- A clear, visual representation of assumptions.
- A systematic method for bias detection and control.
- Guidance for identifying valid adjustment sets (sets of variables needing control).
- A framework for designing causal inference strategies.

Without graphs, assumptions often remain implicit, leading to confusion and potentially invalid inferences.

Causal and ML

Benefits for Machine Learning

Causal graphs can guide the development of more transparent, robust, and reliable machine learning systems.

- **Transparency:** Graphs make the model's assumptions clear and transparent.
- **Bias Control:** They help us identify and control for potential biases in the model.
- **Targeted Interventions:** They show us which parts of the model to modify to achieve a desired outcome.
- **Better Design:** Causal graphs can guide the design of more effective and robust machine learning architectures.

---

10 Causal Graphs and Machine Learning

Causal graphs, with their nodes and edges, provide a powerful analogy for understanding certain machine learning models, especially those that operate on graph-structured data. An adjacency matrix is one way to represent a causal graph (or any directed graph) mathematically.

10.1 Adjacency Matrices

An adjacency matrix is a square matrix where the rows and columns represent the nodes in a graph. The entry at position (i, j) in the matrix indicates whether there is an edge from node i to node j.

- If the entry is 1, there is a directed edge from node i to node j.
- If the entry is 0, there is no directed edge.

For example, consider a simple causal graph: $A \rightarrow B \rightarrow C$. Its adjacency matrix would look like this:

$$
\begin{array}{c|ccc}
 & A & B & C \\
\hline
A & 0 & 1 & 0 \\
B & 0 & 0 & 1 \\
C & 0 & 0 & 0
\end{array}
$$

This matrix precisely encodes the connections: A causes B, and B causes C.

10.2 Graph Neural Networks: Learning on Causal Structures

Graph Neural Networks (GNNs) [2] are a class of machine learning models that directly leverage the adjacency matrix representation of a graph.

- **Message Passing:** GNNs operate by "passing messages" between connected nodes. The adjacency matrix determines which nodes receive messages from which other nodes. This mirrors how causal influence flows along the edges of a causal graph.
- **Causal Inductive Bias:** By their very design, GNNs have a "causal inductive bias." They are naturally suited to learn relationships where the graph structure is important, reflecting underlying causal mechanisms.
- **Examples:**

  - **Social Networks:** A GNN can predict how influence spreads in a social network, where the graph structure represents friendships or follows.
  - **Knowledge Graphs:** GNNs can reason over knowledge graphs, where nodes are entities and edges are relationships (e.g., "part of," "located in").
  - **Molecular Graphs:** GNNs predict chemical properties, where the graph represents atoms and bonds.

10.3 Causal Discovery

Interestingly, some causal discovery [3] algorithms also use the adjacency matrix representation.

- **Learning the Graph:** Algorithms like LiNGAM attempt to learn the adjacency matrix directly (graph structure) from data under certain assumptions. This means the machine learning model is trying to **discover** the causal graph itself.

- **Non-Gaussianity:** LiNGAM, for instance, uses the non-Gaussianity of noise terms to identify the direction of edges in the adjacency matrix.

10.4   Cautions and Extensions

- **GNN Assumption:** Most GNNs and adjacency matrix-based causal discovery [3] methods assume that the underlying graph is a DAG. Handling cycles (feedback loops) is an area of active research.

Beyond Simple Adjacency: Real-world causal relationships and machine learning models can be more complex than what a simple adjacency matrix can capture. Extensions include:

- **Weighted Adjacency Matrices:** To represent the **strength** of causal effects.
- **Dynamic Adjacency Matrices:** To model causal relationships that change over time.

In summary, the adjacency matrix provides a powerful bridge between the visual representation of causal graphs and the mathematical framework of machine learning. Graph Neural Networks [2], in particular, demonstrate how machine learning can effectively learn from and leverage causal structures.

11   Chapter Summary

In this chapter, we introduced causal graphs as a essential tool for representing and reasoning about causal relationships. We learned about nodes, edges, paths, and critical structures like chains, forks, and colliders. Understanding how to construct and read causal graphs lays the groundwork for formal causal inference methods, which we will explore in subsequent chapters.

12   Code

Listing 4.1   Chapter 4: Representing a Causal Graph

```python
import networkx as nx
import matplotlib.pyplot as plt
import numpy as np

# --- 1. Creating a Simple Causal Graph ---
def create_simple_graph():
```

```python
    """Creates and visualizes a simple causal graph:
        Exercise -> Weight Loss."""

    graph = nx.DiGraph()  # Directed Acyclic Graph
    graph.add_edges_from([('Exercise', 'Weight Loss')])

    plt.figure(figsize=(8, 6))
    nx.draw_circular(graph, with_labels=True, node_size
        =2000, node_color='skyblue', font_size=12, arrowsize
        =20)  # Adjust layout for clarity
    plt.title('Simple Causal Graph')
    plt.show()
    return graph


# --- 2. Creating a More Complex Graph ---
def create_complex_graph():
    """Creates and visualizes a more complex causal graph
        with confounding and mediation."""

    graph = nx.DiGraph()
    graph.add_edges_from([('SES', 'Exercise'), ('SES', 'Diet
        '),
                          ('Exercise', 'Weight Loss'), ('
                              Diet', 'Weight Loss')])

    plt.figure(figsize=(8, 6))
    nx.draw_circular(graph, with_labels=True, node_size
        =2000, node_color='lightgreen', font_size=12,
        arrowsize=20)
    plt.title('Complex Causal Graph')
    plt.show()
    return graph


# --- 3. Representing a Graph as an Adjacency Matrix ---
def graph_to_adjacency_matrix(graph):
    """Converts a networkx graph to an adjacency matrix."""

    nodes = list(graph.nodes)
    num_nodes = len(nodes)
    adj_matrix = np.zeros((num_nodes, num_nodes), dtype=int)

    node_to_index = {node: i for i, node in enumerate(nodes)
        }  # Map node to matrix index

    for i in range(num_nodes):
        for j in range(num_nodes):
            if graph.has_edge(nodes[i], nodes[j]):
                adj_matrix[i, j] = 1

    print("\n--- Adjacency Matrix ---")
    print("Nodes:", nodes)
```

```python
    print(adj_matrix)
    return adj_matrix, nodes


# --- 4. Demonstrating a Chain Structure ---
def demonstrate_chain():
    """Illustrates a chain structure and path blocking."""

    graph_chain = nx.DiGraph()
    graph_chain.add_edges_from([('A', 'B'), ('B', 'C')])

    plt.figure(figsize=(6, 4))
    nx.draw_circular(graph_chain, with_labels=True,
        node_size=2000, node_color='lightcoral', font_size
        =12, arrowsize=20)
    plt.title('Chain Structure (A -> B -> C)')
    plt.show()

    print("\n--- Chain Structure ---")
    print("In a chain, B mediates the effect of A on C.")
    print("Conditioning on B blocks the path from A to C.")


# --- 5. Demonstrating a Fork Structure ---
def demonstrate_fork():
    """Illustrates a fork structure and confounding."""

    graph_fork = nx.DiGraph()
    graph_fork.add_edges_from([('B', 'A'), ('B', 'C')])

    plt.figure(figsize=(6, 4))
    nx.draw_circular(graph_fork, with_labels=True, node_size
        =2000, node_color='gold', font_size=12, arrowsize
        =20)
    plt.title('Fork Structure (Confounding)')
    plt.show()

    print("\n--- Fork Structure ---")
    print("In a fork, B is a common cause of A and C (
        confounder).")
    print("Not controlling for B can create a spurious
        association between A and C.")


# --- 6. Demonstrating a Collider Structure ---
def demonstrate_collider():
    """Illustrates a collider structure and collider bias.
        """

    graph_collider = nx.DiGraph()
    graph_collider.add_edges_from([('A', 'B'), ('C', 'B')])

    plt.figure(figsize=(6, 4))
```

```
    nx.draw_circular(graph_collider, with_labels=True,
        node_size=2000, node_color='lightblue', font_size
        =12, arrowsize=20)
    plt.title('Collider Structure')
    plt.show()

    print("\n--- Collider Structure ---")
    print("In a collider, B is caused by both A and C.")
    print("Conditioning on B can open a non-causal path
        between A and C, creating bias.")


# --- 7. Main Execution ---
if __name__=="__main__":
    simple_graph = create_simple_graph()
    complex_graph = create_complex_graph()

    adj_matrix, nodes = graph_to_adjacency_matrix(
        complex_graph)

    demonstrate_chain()
    demonstrate_fork()
    demonstrate_collider()
```

13 Exercises

1. Draw a simple Directed Acyclic Graph (DAG) showing confounding.
2. What does it mean for a graph to be acyclic?
3. Explain the role of backdoor paths in causal inference.
4. How does blocking backdoor paths help in estimating causal effects?

References

1. Pearl, Judea. 2009. *Causality: Models, Reasoning, and Inference.* 2nd ed. Cambridge University Press.
2. Scarselli, Franco, et al. 2009. The graph neural network model. *IEEE Transactions on Neural Networks* 20 (1): 61–80.
3. Spirtes, Peter, Clark N Glymour, and Richard Scheines. 2000. *Causation, Prediction, and Search.* MIT Press.

Interventions and Counterfactuals

1 Introduction to Interventions

Many scientific and practical questions we care about are causal. We often want to know whether two variables are related and what happens if we actively change one of them.

For example, imagine a new medication is being developed. It is not enough to observe that patients who choose the medication tend to recover faster. Perhaps healthier patients are more likely to opt for the drug. Instead, we would like to administer the drug and see if recovery rates improve actively.

In observational studies, we passively observe variables without any control. Variables evolve according to their natural mechanisms, including hidden biases and confounding factors. However, when we perform an **intervention**, we forcibly change the value of a variable, independently of its usual causes.

In causal inference, interventions are formalized through the *do-operator* [1], denoted as $\mathrm{do}(T = t)$. This operation cuts off any natural causes of the variable T and assigns it a fixed value t.

> **Definition (Intervention):** An intervention [1] is an external manipulation that sets a variable to a specific value, independently of the variable's natural causal parents.

This core idea distinguishes causal inference from traditional statistical learning.

In Fig. 1, smoking (T) is naturally influenced by genetic factors (Z), and both impact lung cancer (Y). Observational data captures correlations along all these paths. However, if we intervene to set smoking status manually, we break the influence of genetics on smoking.

© The Author(s), under exclusive license to Springer Nature Switzerland AG 2025
D. Rajamanickam, *Causal Inference for Machine Learning Engineers*,
https://doi.org/10.1007/978-3-031-99680-1_5

Fig. 1 Causal graph:
Genetics influence smoking,
influencing lung cancer

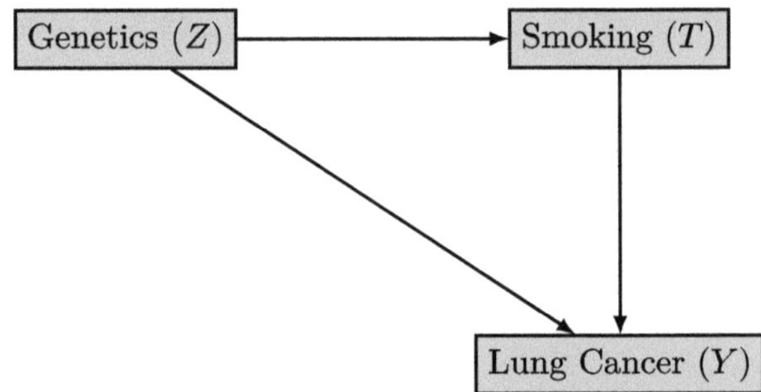

> **Causal and ML**
>
> ***Interventions and Actions in ML***
>
> Interventions in causal inference provide a framework for understanding the
> effect of actions taken by machine learning agents or systems.
>
> - **Beyond Passive Observation:** Machine learning models typically
>   **observe** data. Interventions are about actively **changing** the data and see-
>   ing what happens. It's like going from watching a video to controlling the
>   game.
> - **Manipulating Inputs:** In machine learning, interventions are like directly
>   manipulating the input features. What happens if we increase the contrast
>   of an image? What if we remove a word from a sentence?
> - **Reinforcement Learning as Intervention:** Reinforcement learning is all
>   about interventions. The agent **acts** in the environment, which changes
>   the environment's state. Causal inference helps us understand the **effect**
>   of those actions.

2   Observational Versus Interventional Distributions

The distinction between observational and interventional distributions lies at the
heart of causal inference.

The observational distribution $P(Y \mid T)$ represents the probability of the outcome
Y given that T is observed naturally. Here, T could have been influenced by hidden
confounders.

In contrast, the interventional distribution $P(Y \mid \mathrm{do}(T))$ represents the probability
of Y under an external manipulation that sets T to a specific value, bypassing any of
its natural causes.

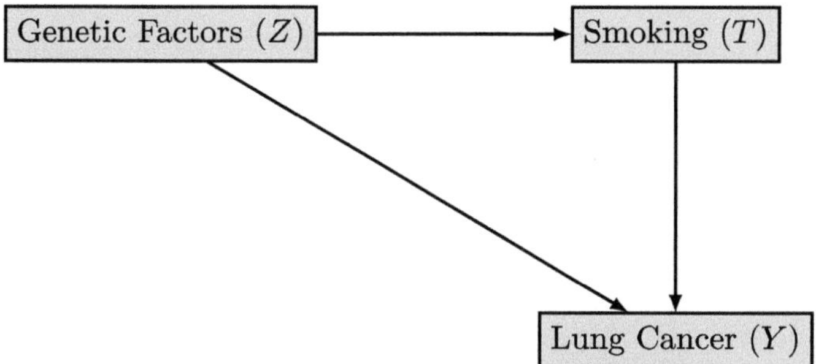

Fig. 2 Observational system: natural pathways from Z to T and Y exist

Fig. 3 Interventional
system: Intervention breaks
the natural path $Z \to T$

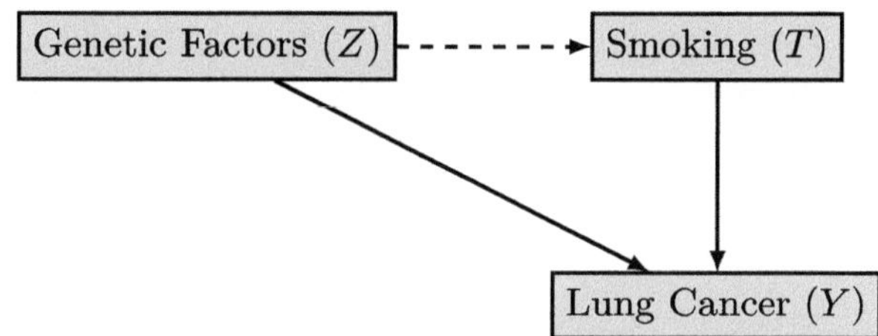

Mathematically, the difference is clear. Observationally:

$$P(Y \mid T) = \sum_Z P(Y \mid T, Z) P(Z \mid T),$$

While interventionally:

$$P(Y \mid \mathrm{do}(T)) = \sum_Z P(Y \mid T, Z) P(Z).$$

The key change is that $P(Z)$ appears instead of $P(Z \mid T)$. After an intervention, Z and T are no longer statistically dependent (Fig. 2).

In Fig. 3, the dashed arrow shows that the causal influence of Z on T is removed after intervention.

Causal and ML

Two Ways of Seeing the World

The difference between observational and interventional distributions highlights the limitations of purely correlational machine learning models for decision-making.

The Model's Natural View: The observational distribution ($P(Y|T)$) is what the machine learning model learns from the data it's trained on. It's the model's "natural" view of the world.

The "What If" View: The interventional distribution ($P(Y|do(T))$) is what we get when we **force** the treatment. It's the model's view of a **modified** world.

The Difference Matters: If there's confounding, these two views will be different. If we start intervening, the model trained on observational data will make wrong predictions.

---

3   Understanding the Do-Operator

The do-operator allows us to describe interventions in a causal model mathematically.

We compute $P(Y \mid T = t)$ by conditioning on an observed value in ordinary probability. However, in causality, we distinguish this from $P(Y \mid do(T = t))$, where T is forcibly set by intervention.

The do-operator has two essential effects on the causal graph:

- All arrows pointing into T are cut.
- T is assigned the value t.

This procedure isolates the true causal effect of T on Y.

In formal terms:

$$P(Y \mid do(T = t)) = \sum_Z P(Y \mid T = t, Z) P(Z),$$

which differs from merely observing T and Y.

Causal and ML

The Do-Operator: The Intervention Switch

The do-operator formalizes the concept of manipulating input variables, which is essential for understanding the causal effect of feature changes on a model's output.

Cutting the Wires: The do-operator $(do(T = t))$ is like a switch that cuts off the natural causes of the treatment variable. We're not letting T be what it wants; we're setting it ourselves.

Breaking Dependencies: This "cutting the wires" changes the relationships in the data. Variables correlated with T might no longer be correlated after we do(T).

Example (Text Generation):

- Observational: Model learns that negative sentiment is correlated with certain words.
- Interventional: If we **force** the model to use those words, does it **still** generate negative sentiment?

---

4 Causal Effects via Intervention

Having established the meaning of intervention, we now define causal effects.

The causal effect of a treatment T on an outcome Y is the change in the distribution of Y caused by setting T to different values through intervention.

For binary treatments, one quantity of particular interest is the **Average Treatment Effect (ATE)**:

$$\text{ATE} = \mathbb{E}[Y \mid do(T = 1)] - \mathbb{E}[Y \mid do(T = 0)].$$

This measures the expected difference in outcomes if everyone in the population were treated versus if no one were treated.

For instance, in Fig. 4, family background (Z) influences both college attendance (T) and income (Y). To estimate the causal effect of education on income, we must account for Z properly.

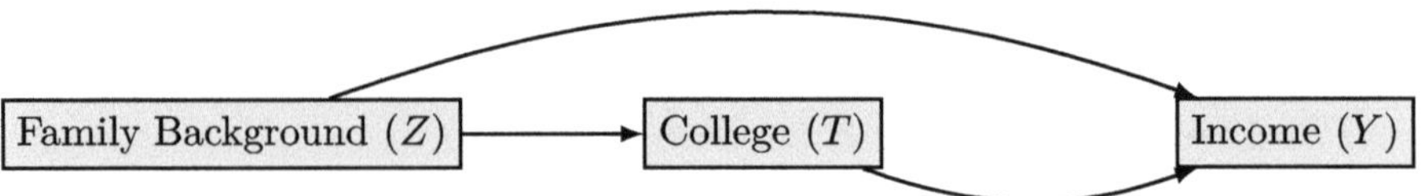

Fig. 4 Causal graph for education (T) and income (Y), with family background (Z) as confounder

---

Causal and ML

Measuring the Impact of Change

Estimating causal effects via intervention is crucial for evaluating the impact of changes to machine learning systems, such as deploying a new model or modifying a recommendation algorithm.

- **The Intervention's Footprint:** The causal effect is how much the intervention **changes** the outcome. It's the "footprint" our action leaves on the world.
- **Average Change:** The Average Treatment Effect (ATE) is the average "footprint" across a whole population. It's like the average effect of a new website design on user engagement.
- **Example (Personalized Recommendations):**

  - We want to know: Does showing a user **this specific** movie **cause** them to watch it?
  - The ATE tells us the average effect across all users.

---

5   Counterfactuals: Imagining Alternate Worlds

Interventions predict what happens under manipulation, but counterfactuals ask about alternative realities: What would have happened if something different had occurred? A counterfactual question might be:

"If this patient had received treatment, would they have survived?"

Formally, for a unit with covariates X, treatment T, and outcome Y, we define potential outcomes:

$$Y(1) = \text{Outcome if treated},$$

$$Y(0) = \text{Outcome if untreated}.$$

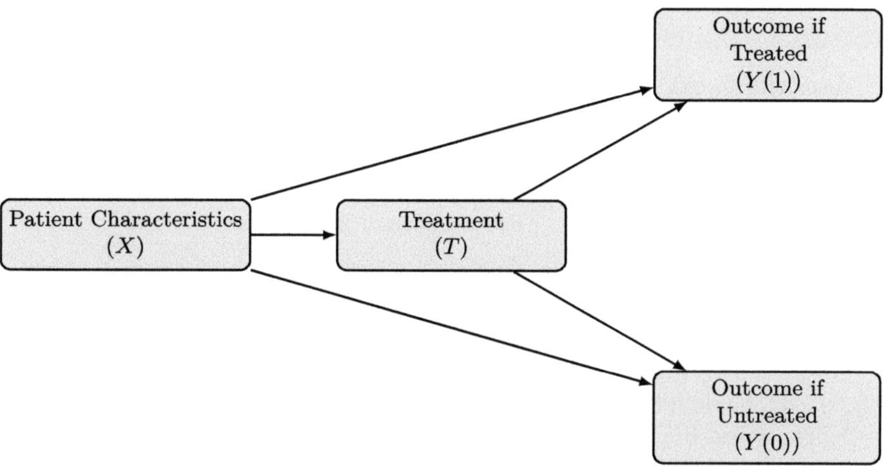

Fig. 5 Causal diagram showing counterfactual structure: only one of $Y(1)$ or $Y(0)$ is observed

We only observe one of these outcomes in reality. The other remains a hypothetical, requiring modeling assumptions to infer (Fig. 5).

Counterfactual reasoning is essential in policy evaluation, fairness in AI, and legal responsibility analysis.

Key Concepts Summarized:

$$P(Y \mid do(T = t)) \quad \text{(intervention probability)}$$

$$\text{ATE} = \mathbb{E}[Y \mid do(T = 1)] - \mathbb{E}[Y \mid do(T = 0)]$$

$$\text{Counterfactual outcomes: } Y(1), Y(0)$$

These ideas set the stage for formal causal analysis, which we will deepen further in subsequent chapters.

> ### Causal and ML
>
> ### *Counterfactuals: The "What Ifs" of ML*
>
> Counterfactuals provide a powerful tool for explaining machine learning decisions and assessing fairness by asking questions like 'What would the model have predicted if the input had been different?'
>
> **Beyond Prediction:** Machine learning usually asks, "What **will** happen?" Counterfactuals ask, "What *would have* happened if things were different?"
> **Debugging Model Decisions:** Counterfactuals are incredibly useful for understanding **why** a model made a certain decision. "Why was this loan denied? Would it have been approved if the income was higher?"
> **Fairness and Bias:** Counterfactuals help us detect bias. "Would this person have gotten the job if they had a different name?"
> **Example (Medical Diagnosis):**
>
> - Model says: "This patient has a high risk of disease."
> - Counterfactual: "Would the risk have been lower if they had received treatment earlier?"

6   Summary

In this chapter, we introduced two core ideas in causal inference: **interventions** and **counterfactuals**.

An **intervention** refers to an external action that forces a variable to take a specific value, breaking the usual relationships in the system. We formalized interventions using the **do-operator** $do(T = t)$, which explicitly distinguishes observing a variable from actively setting it. While $P(Y \mid T)$ measures the probability of an outcome given observation, $P(Y \mid do(T))$ (The do-operator, introduced by Pearl [1], formalizes interventions by specifying external actions that set variables to fixed values.) measures the outcome under intervention, a crucial difference when inferring causality.

We also discussed the concept of **counterfactual outcomes**. A counterfactual asks: "What would have happened if things had been different?" For any individual, counterfactual reasoning imagines the outcome under a treatment they did not receive. Counterfactuals allow deeper insights into causality by considering alternate realities and are key to answering individual-level causal questions.

Importantly, we emphasized that counterfactuals are usually **unobservable**—we only ever see one factual outcome for each individual. Despite this, causal models can estimate counterfactual quantities by making assumptions about the data-generating process.

Through formalism and practical examples, we saw that distinguishing between observation and intervention is critical for correctly answering causal questions. That counterfactual reasoning lies at the heart of causal decision-making.

In the next chapter, we will build on these ideas by introducing **do-calculus** [1], a robust set of rules that allow us to manipulate and identify causal effects from observational data under certain conditions.

7 Code

Listing 5.1 Chapter 5: Illustrating the Intervention and Counterfactuals

```python
import numpy as np
import pandas as pd
import matplotlib.pyplot as plt
import statsmodels.formula.api as sm  # For regression

# --- 1. Simulate Data with Confounder ---
def simulate_confounded_data(n=1000, effect_of_z=1.0,
    effect_of_t=2.0):
    """Simulates data with a confounder Z affecting both T
        and Y."""

    Z = np.random.normal(0, 1, n)  # Confounder
    T = 0.5 * Z + np.random.normal(0, 1, n)  # Treatment
        depends on Z
    Y = effect_of_t * T + effect_of_z * Z + np.random.normal
        (0, 1, n)  # Outcome depends on both
    return pd.DataFrame({'Z': Z, 'T': T, 'Y': Y})


# --- 2. Observational Analysis: Y vs. T ---
def observational_analysis(data):
    """Calculates and visualizes the observational
        relationship between T and Y."""

    print("\n--- Observational Analysis (Y vs. T) ---")
    print(data.groupby('T')['Y'].mean())  # Grouped means

    plt.figure(figsize=(6, 4))
    sns.regplot(x='T', y='Y', data=data, ci=95)  #
        Regression line with CI
    plt.xlabel('Treatment (T)')
    plt.ylabel('Outcome (Y)')
    plt.title('Observational: Y vs. T')
    plt.show()

    model = sm.ols('Y ~ T', data=data).fit()
```

```python
    print(model.summary())   # Regression summary (optional)


# --- 3. Interventional Analysis: do(T) ---
def interventional_analysis(data, t_value=0):
    """Simulates the do-operator by setting T to a fixed
        value and analyzing Y."""

    print(f"\n--- Interventional Analysis: do(T = {t_value})
        ---")

    data_do_t = data.copy()
    data_do_t['T'] = t_value  # Intervention: Set T to '
        t_value'

    print(f"Mean Outcome when do(T = {t_value}): {data_do_t
        ['Y'].mean():.2f}")

    plt.figure(figsize=(6, 4))
    plt.hist(data_do_t['Y'], bins=20, alpha=0.7)
    plt.xlabel('Outcome (Y)')
    plt.ylabel('Frequency')
    plt.title(f'Interventional: Distribution of Y when do(T
        = {t_value})')
    plt.show()

    model_do_t = sm.ols('Y ~ 1', data=data_do_t).fit()   #
        Regression on constant
    print(model_do_t.summary())  # Summary (optional)


# --- 4. Comparing Observational and Interventional ---
def compare_distributions(data, data_do_0, data_do_1):
    """Compares the distributions of Y under observational
        and interventional settings."""

    plt.figure(figsize=(8, 5))

    sns.kdeplot(data['Y'], label='Observational', fill=True,
        alpha=0.3)
    sns.kdeplot(data_do_0['Y'], label='do(T=0)', fill=True,
        alpha=0.3)
    sns.kdeplot(data_do_1['Y'], label='do(T=1)', fill=True,
        alpha=0.3)

    plt.xlabel('Outcome (Y)')
    plt.ylabel('Density')
    plt.title('Comparison: Observational vs. Interventional'
        )
    plt.legend()
    plt.show()
```

```
# --- 5. Main Execution ---
if __name__=="__main__":
    np.random.seed(123)
    data = simulate_confounded_data(effect_of_t=2.0)   #
        Adjust the 'true' effect of T

    observational_analysis(data)
    interventional_analysis(data, t_value=0)
    interventional_analysis(data, t_value=1)

    data_do_0 = data.copy()
    data_do_0['T'] = 0
    data_do_1 = data.copy()
    data_do_1['T'] = 1
    compare_distributions(data, data.copy(), data_do_0,
        data_do_1)

    print("\n--- Explanation ---")
    print("This code simulates data with a confounder 'Z'
        affecting both treatment 'T' and outcome 'Y'.")
    print("It then demonstrates the difference between:")
    print("1.  Observational analysis:  Analyzing the
        relationship between 'T' and 'Y' as observed in the
        data.")
    print("2.  Interventional analysis:  Simulating the 'do
        (T)' operation by setting 'T' to a fixed value and
        observing the resulting distribution of 'Y'.")
    print("The comparison of distributions highlights how
        interventions can change the relationship between
        variables compared to passive observation.")
```

8   Exercises

1. Explain the concept of an intervention using the do-operator.
2. What is the difference between $P(Y \mid T)$ and $P(Y \mid do(T))$?
3. Define a counterfactual outcome and explain why it is important.
4. Give a practical example where counterfactual reasoning changes a decision.

Reference

1. Pearl, Judea. 2009. *Causality: Models, Reasoning, and Inference*. 2nd ed. Cambridge University Press.

Introduction to Do-Calculus

1 Why Do-Calculus?

Observational data alone often falls short in understanding causal relationships. Although it is straightforward to observe correlations between variables, distinguishing causation from mere association is a far deeper and more complex challenge. Confounding variables often create spurious relationships that traditional statistical methods cannot untangle.

The need for a formal system to reason about interventions led to the development of **do-calculus** [1] by Judea Pearl [1] and his collaborators. Do-calculus provides a rigorous mathematical framework for transforming expressions involving interventions into expressions involving only observed quantities, whenever possible.

Without do-calculus [1], the practitioner is often limited to guessing which variables to adjust for, based on intuition. Do-calculus systematically tells us when and how causal effects can be identified, using only the observed data and the structure of the causal graph.

The brilliance of do-calculus [1] lies in its providing a **symbolic manipulation system** for interventions, much like algebra provides a system for manipulating equations.

> **Key Insight**
>
> Do-calculus [1] enables us to answer interventional queries using observational data, provided we know the causal structure.

Thus, do-calculus [1] acts as the bridge between observational studies and interventional reasoning, which is the very heart of causal inference [1].

D. Rajamanickam, *Causal Inference for Machine Learning Engineers*,
https://doi.org/10.1007/978-3-031-99680-1_6

> ### Causal and ML
>
> ## *Do-Calculus: The Algebra of Interventions*
>
> Do-calculus [1] provides a formal system for reasoning about interventions, enabling machine learning engineers to design more robust and reliable systems.
>
> - **Beyond Observational Data:** Machine learning models learn from patterns in observed data. Do-calculus [1] provides a way to reason about what happens when we actively **change** the data, not just passively watch it.
> - **Manipulating Interventions:** Do-calculus [1] is like an algebra for interventions. It gives us rules to manipulate expressions with the do() operator, similar to how we manipulate equations.
> - **Causal Effect Identification:** Do-calculus [1] helps us systematically determine whether we can estimate a causal effect from observational data alone or need experimental interventions.
> - **Analogy: Model Surgery:** Imagine a machine learning model as a complex circuit. Do-calculus [1] helps us figure out which "wires" to cut or reroute to isolate the effect of a specific input on the output.

2 The Do-Operator Revisited

Before diving into the rules of do-calculus [1], let us first revisit the idea of the **do-operator**.

When we write $do(T = t)$, we conceptualize an intervention that forcibly sets the treatment variable T to a specific value t, irrespective of its usual causes. This operation is depicted in causal diagrams by removing all arrows pointing into T.

Formally, if G is a causal graph, then $G_{do(T)}$ is the graph obtained by deleting all incoming edges into T (Fig. 1).

In Fig. 2, the incoming edge $Z \rightarrow T$ is dashed to indicate that it has been severed by the intervention $do(T)$.

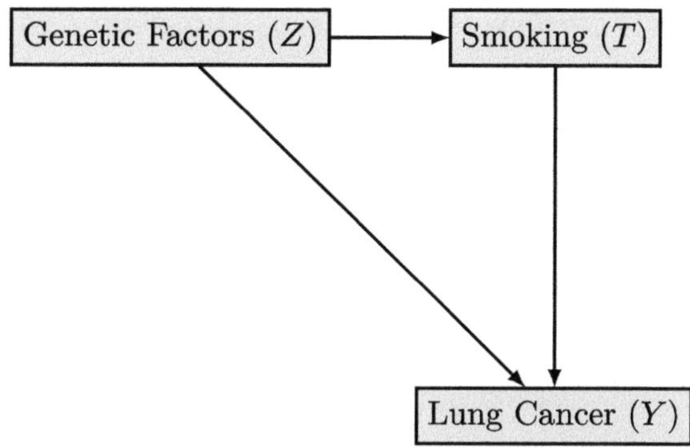

Fig. 1 Causal graph before intervention—$Z \to T$ and $Z \to Y$

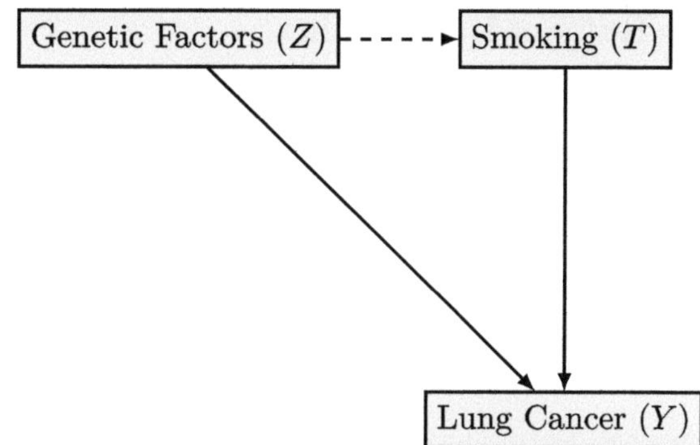

Fig. 2 Causal graph after intervention—incoming edge into T removed

---

Causal and ML

The do() Operator: The Intervention Switch

The do-operator helps distinguish between observing data correlations and actively manipulating variables, a crucial distinction for causal machine learning.

- **Forced Values:** The do(T=t) operator is like forcing a specific input feature (T) to take on a certain value (t), regardless of what its normal value would be.
- **Graph Modification:** Graphically, do(T=t) is represented as removing all arrows pointing into the node representing T in the causal graph. This symbolizes breaking the natural causes of T.
- **Observational vs. Interventional:** P(Y|T=t) is what we observe in the data. P(Y|do(T=t)) is what happens when we **force** T to be t. These are generally different!
- **Analogy: Model Input Control:** In a machine learning model, do(T=t) is like taking direct control of a specific input neuron and setting its activation to a fixed value.

2.1　Observational Versus Interventional Distributions

The difference between observing $T = t$ and doing $T = t$ is profound:

$$P(Y \mid T = t) \neq P(Y \mid \mathrm{do}(T = t))$$

Observational distributions are influenced by confounders, while interventional distributions represent the outcome under controlled manipulation.

Thus, learning how to convert between them is critical, and this is precisely what do-calculus allows.

3　Rules of Do-Calculus

Pearl's [1] do-calculus consists of three rules that manipulate probabilities involving $\mathrm{do}(\cdot)$ operations. These rules tell us when and how to remove or modify a do-operator in an expression using the graph's structure.

3.1　Preliminaries

Consider a causal graph G with nodes representing random variables and directed edges representing causal relationships. Suppose X, Y, Z, and W are disjoint sets of variables in G.

The rules of do-calculus tell us how to transform probabilities involving $\mathrm{do}(\cdot)$ into expressions without do, whenever possible.

Each rule involves conditions about **d-separation**—a graphical property that indicates conditional independence.

Do-calculus, a set of three rules proposed by Pearl [1], enables the identification of causal effects from observational data under certain graphical conditions.

3.2　Rule 1: Insertion/Deletion of Observations

Statement:

$$P(Y \mid \mathrm{do}(X), Z, W) = P(Y \mid \mathrm{do}(X), W) \quad \text{if } Y \perp\!\!\!\perp Z \mid X, W \text{ in } G_{\overline{X}}$$

Here, $G_{\overline{X}}$ is the graph obtained from G by deleting all incoming edges into X.

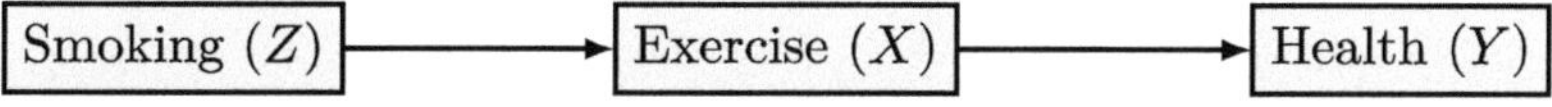

Fig. 3 Graph where Z influences X, which influences Y

Interpretation:

If Y is independent of Z given X and W in the graph where we have intervened on X, then conditioning on Z becomes redundant. Thus, we can remove Z from the conditioning set.

> **Key Idea for Rule 1**
>
> Observing Z after intervening on X provides no additional information about Y if Z is d-separated from Y given X and W.

Graphical Example:

Suppose X affects Y, and Z affects X but not Y directly. After cutting incoming edges into X, Z becomes irrelevant for predicting Y (Fig. 3).

After doing X, the influence of Z is cut off.

> **Causal and ML**
>
> **Rule 1: Redundant Information**
>
> **Irrelevant Inputs:** Rule 1 says that if observing another variable (Z) doesn't give us any new information about the outcome (Y) after we've intervened on X, then we can ignore Z.
>
> **Model Simplification:** This is like simplifying a machine learning model by removing an input that doesn't contribute to the prediction after considering other inputs.
>
> **Analogy: Feature Selection:** Rule 1 helps us select redundant features that can be safely discarded.

3.3 Rule 2: Action/Observation Exchange

Statement:

$$P(Y \mid \mathrm{do}(X), \mathrm{do}(Z), W) = P(Y \mid \mathrm{do}(X), Z, W) \quad \text{if } Y \perp\!\!\!\perp Z \mid X, W \text{ in } G_{\overline{X}, \underline{Z}}$$

Here:

- $G_{\overline{X},\underline{Z}}$ is the graph where incoming edges into X are cut (due to do(X)), and outgoing edges from Z are cut (due to do(Z)).

Interpretation:

If, after intervening on X, observing Z is just as good as intervening on Z, then we can replace do(Z) with simply conditioning on Z.

> **Key Idea for Rule 2**
>
> Under certain conditions, an intervention can be replaced by an observation.

Graphical Example:

Imagine a case where Z only affects Y through paths already blocked by conditioning on W and X. Then intervening on Z has no additional effect beyond observing it.

> **Causal and ML**
>
> **Rule 2: Intervention as Observation**
>
> **Equivalent Actions:** Rule 2 states that intervening on a variable (Z) has the same effect as simply observing it under certain conditions.
> **Model Equivalence:** This is like finding two different ways to achieve the same result in a machine learning model: actively setting a neuron's value or letting it take its natural value.
> **Analogy: Model Optimization:*** Rule 2 can sometimes help us optimize a model by replacing an expensive intervention with a cheaper observation.

3.4 Rule 3: Insertion/Deletion of Actions

Statement:

$$P(Y \mid \text{do}(X), \text{do}(Z), W) = P(Y \mid \text{do}(X), W) \quad \text{if } Y \perp\!\!\!\perp Z \mid X, W \text{ in } G_{\overline{X},\overline{Z}}$$

Here:

- $G_{\overline{X},\overline{Z}}$ is the graph where all incoming edges into X and Z have been removed.

Interpretation:

If Y is independent of Z given X and W after cutting edges into X and Z, then the intervention on Z cannot affect Y and can be deleted.

> **Key Idea for Rule 3**
>
> It can be ignored if an intervention has no causal impact on the outcome.

Graphical Example:

If Z only affects X and not Y directly, then after doing X, Z becomes irrelevant to Y.

> **Causal and ML**
>
> **Rule 3: Irrelevant Interventions**
>
> **Useless Actions:** Rule 3 says that if intervening on a variable (Z) does not affect the outcome (Y), then we can ignore that intervention.
> **Model Pruning:** This is like pruning a machine learning model by removing network parts that don't affect the output.
> **Analogy: Input Importance:** Rule 3 helps us identify unimportant input features that have no causal effect on the prediction.

4 Simple Example of Do-Calculus

Let us consider a concrete example:

We want to estimate $P(Y \mid \mathrm{do}(T))$—the causal effect of education on income (Fig. 4).
- In observational data, Z confounds the relationship between T and Y.
- However, by adjusting for Z, we can block the backdoor path.

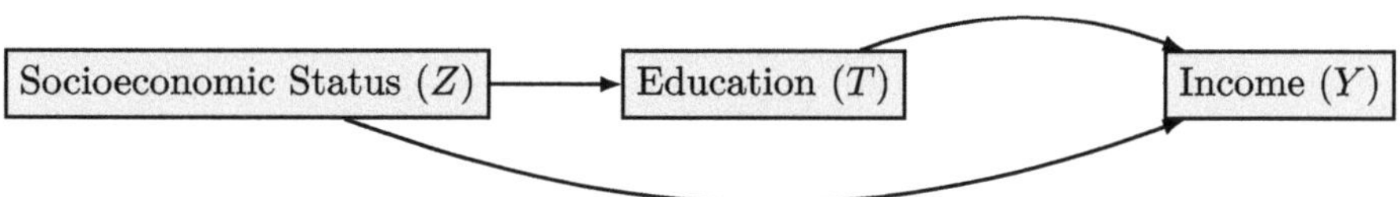

Fig. 4 Education and income, confounded by socioeconomic status

Thus:

$$P(Y \mid \mathrm{do}(T)) = \sum_z P(Y \mid T, Z = z)P(Z = z)$$

which is known as the **backdoor adjustment formula**.

This shows the power of combining causal graphs, the do-operator, and adjustment.

Causal and ML

Do-Calculus in Action: Education and Income

Do-calculus [1] can guide the development of machine learning models less susceptible to confounding bias.

Confounding in ML: This example illustrates how confounding (socioeconomic status affecting education and income) can mislead a machine learning model.

Backdoor Adjustment: Do-calculus tells us that we can correct this confounding by "adjusting" for socioeconomic status, which is like adding it as an input to the model.

Analogy: Feature Control: Do-calculus helps us decide which features to control for in our machine learning model to get unbiased estimates of feature importance.

---

5 Backdoor Criterion

The **backdoor criterion** [1] gives a graphical condition for when adjustment is sufficient.

A set of variables Z satisfies the backdoor criterion relative to (T, Y) if:

1. No node in Z is a descendant of T, and

2. Z blocks every path between T and Y that contains an arrow into T.

If Z satisfies the backdoor criterion, then:

$$P(Y \mid \mathrm{do}(T)) = \sum_z P(Y \mid T, Z = z)P(Z = z)$$

> **Causal and ML**
>
> ### *Backdoor Criterion: The Recipe for Unbiasedness*
>
> The backdoor criterion [1] provides a principled way to select features for machine learning models, ensuring that the model learns causal relationships.
>
> - **Graphical Recipe:** The backdoor criterion provides a graphical "recipe" for choosing which variables to adjust for to get an unbiased causal effect estimate.
> - **Feature Selection Guidance:** This criterion offers a principled way to select features for a machine learning model, ensuring we control for the right variables to avoid bias.
> - **Analogy: Data Preprocessing:** Backdoor adjustment is like a sophisticated form of data preprocessing that removes the influence of confounding variables.

6 Frontdoor Criterion

Sometimes, backdoor adjustment is impossible. The **frontdoor criterion** [1] offers an alternative:

A set of variables M satisfies the frontdoor criterion relative to (T, Y) if:

1. All directed paths from T to Y pass through M.

2. There are no unblocked backdoor paths from T to M.

3. All backdoor paths from M to Y are blocked by T.

Under these conditions, the causal effect can be computed via:

$$P(Y \mid \mathrm{do}(T)) = \sum_m P(M = m \mid T) \sum_{t'} P(Y \mid M = m, T = t') P(T = t')$$

Frontdoor adjustment uses mediators to recover causal effects even when backdoor paths are unblocked.

Frontdoor Criterion: An Alternative Path

The frontdoor criterion [1] can be used to understand the mechanisms by which machine learning models make predictions, improving their interpretability.

- **Mediator as Intermediary:** The frontdoor criterion offers an alternative approach when we can't directly measure all confounders. It uses a mediator variable based on the causal path between treatment and outcome.
- **Model Decomposition:** This is like decomposing a machine learning model into smaller sub-models, where the mediator acts as an intermediary layer.
- **Analogy: Step-by-Step Explanation:** Frontdoor adjustment helps us understand the step-by-step mechanism by which the treatment affects the outcome.

7 Chapter Summary

In this chapter, we introduced **do-calculus** [1], a robust mathematical framework that allows us to reason about interventions using observational data.

The need for do-calculus arises because performing actual interventions in many real-world situations is difficult, expensive, or impossible. Instead, we have observational data and want to infer causal effects from it. Do-calculus provides a formal set of rules that enable us to manipulate expressions involving the do-operator $do(T)$ and connect interventional distributions to purely observational ones.

We explored the three main rules of do-calculus, each allowing us to move or eliminate the do-operator under specific graphical conditions related to d-separation in causal graphs. These rules allow us to identify when and how causal effects are estimable from observed data alone.

Importantly, we learned that not all causal effects can be identified without interventions; do-calculus tells us precisely when this is possible and how to proceed.

Through examples, we saw how applying do-calculus systematically can unlock answers to complex causal questions, guiding us from raw observational data to causal conclusions.

In the next chapter, we will delve deeper into specific graphical criteria, the backdoor and frontdoor criteria, which provide practical conditions for identifying causal effects from observational data.

8 Code

Listing 6.1 Chapter 6: Focusing on demonstrating the rules of do-calculus with simulations

```python
import numpy as np
import pandas as pd
import statsmodels.formula.api as sm  # For regression

# --- 1. Simulate Data for Backdoor Criterion Example ---
def simulate_backdoor_data(n=1000, effect_t_y=2, effect_z_t
    =0.5, effect_z_y=1):
    """Simulates data to demonstrate the backdoor criterion.
        """

    Z = np.random.normal(0, 1, n)  # Confounder
    T = effect_z_t * Z + np.random.normal(0, 1, n)  #
        Treatment
    Y = effect_t_y * T + effect_z_y * Z + np.random.normal
        (0, 1, n)  # Outcome
    return pd.DataFrame({'Z': Z, 'T': T, 'Y': Y})


# --- 2. Demonstrate Naive (Biased) Effect Estimation ---
def naive_effect_estimation(data):
    """Estimates the effect of T on Y without adjusting for
        Z (biased)."""

    model = sm.ols('Y ~ T', data=data).fit()
    ate_naive = model.params['T']
    print("\n--- Naive Effect Estimation (Biased) ---")
    print(f"Naive ATE: {ate_naive:.2f}")
    print("This estimate is biased because it doesn't
        account for the confounder Z.")
    return ate_naive


# --- 3. Demonstrate Backdoor Adjustment ---
def backdoor_adjustment(data):
    """Estimates the effect of T on Y using backdoor
        adjustment (adjusting for Z)."""

    model = sm.ols('Y ~ T + Z', data=data).fit()
    ate_adjusted = model.params['T']
    print("\n--- Backdoor Adjustment ---")
    print(f"Adjusted ATE: {ate_adjusted:.2f}")
    print("This estimate is (unbiased) because it adjusts
        for the confounder Z, satisfying the backdoor
        criterion.")
    return ate_adjusted, model


# --- 4. Illustrate Do-Calculus Rule 3 (Simplified) ---
def illustrate_do_calculus_rule3(data):
```

```python
    """Illustrates a simplified version of do-calculus Rule
        3 using the simulated data."""

    print("\n--- Do-Calculus Rule 3 Illustration (Simplified
        ) ---")
    print("Rule 3 states that if a variable Z is independent
         of Y given X after intervening on X, then do(Z) is
        irrelevant for predicting Y.")
    print("In our case, if we were to simulate 'do(T)', the
        relationship between Z and Y would change.")
    print("However, in this simplified simulation, we'll
        show the effect of adjusting vs. not adjusting.")

    # This is NOT a true 'do()' simulation, but it shows the
         effect of Z
    model_without_do_z = sm.ols('Y ~ T', data=data).fit()
    effect_without_do_z = model_without_do_z.params['T']
    print(f"\nEffect without adjusting for Z: {
        effect_without_do_z:.2f}")

    model_with_do_z = sm.ols('Y ~ T + Z', data=data).fit()
    effect_with_do_z = model_with_do_z.params['T']
    print(f"Effect after adjusting for Z (analogous to 'do(T
        )'): {effect_with_do_z:.2f}")   # Analogous to do(T)

    print("\nIn this simplified example, adjusting for Z is
        similar to what 'do(T)' would achieve (in terms of
        removing confounding).")


# --- 5. Main Execution ---
if __name__=="__main__":
    np.random.seed(42)   # Consistent seed
    data = simulate_backdoor_data(effect_t_y=2, effect_z_t
        =0.5, effect_z_y=1)   # Control causal strengths

    naive_effect = naive_effect_estimation(data)
    adjusted_effect, regression_model = backdoor_adjustment(
        data)
    illustrate_do_calculus_rule3(data)   # Simplified
        illustration

    print("\n--- Summary ---")
    print("This code demonstrates how confounding biases the
         estimation of the effect of T on Y.")
    print("It shows how backdoor adjustment (analogous to do
        -calculus) can correct for this bias.")
    print("Specifically, it illustrates (in a simplified way
        ) how do-calculus helps us identify and remove the
        influence of Z when estimating the effect of T on Y.
        ")
```

9 Exercises

1. What is do-calculus and why is it important in causal inference?
2. Explain the intuition behind one of the rules of do-calculus.
3. Provide an example where do-calculus allows identification of a causal effect.
4. Why do we need formal rules for moving between observational and interventional distributions?

Reference

1. Pearl, Judea. 2009. *Causality: Models, Reasoning, and Inference.* 2nd ed. Cambridge University Press.

Backdoor and Frontdoor Criteria

1 Introduction

In causal inference [1], a central goal is to estimate the causal effect of a treatment variable T on an outcome variable Y. Observational data, where treatment assignment is not controlled, often suffers from confounding, leading to biased estimates of this effect [cite: 2, 3]. This chapter introduces two key criteria—the backdoor criterion [1] and the frontdoor criterion [1] which, under certain assumptions, allow us to identify causal effects from observational data.

Causal and ML

Causal Identification in Machine Learning

Backdoor and frontdoor criteria provide a causal framework for feature selection and model design in machine learning, ensuring that models learn genuine causal relationships.

- **The Bias Problem**: Machine learning models can learn biased relationships from observational data, leading to poor generalization and unfair predictions.
- **Causal Solutions**: The backdoor and frontdoor criteria provide principled ways to select features and structure models to mitigate this bias.
- **Analogy: Model Debugging**: These criteria are like debugging tools that help us identify and remove the sources of bias in our models.

© The Author(s), under exclusive license to Springer Nature Switzerland AG 2025
D. Rajamanickam, *Causal Inference for Machine Learning Engineers*,
https://doi.org/10.1007/978-3-031-99680-1_7

2   The Challenge of Confounding

Confounding occurs when a third variable, Z, influences both the treatment T and the outcome Y, creating a spurious association between them. Naive comparisons of Y for different values of T will then reflect not only the causal effect of T on Y but also the influence of Z. To obtain unbiased estimates, we must adjust for Z.

> **Causal and ML**
>
> ### *Confounding: The Hidden Influence*
>
> Confounding is a major source of misleading patterns in machine learning data, and causal inference techniques are essential for building robust models.
>
> - **Spurious Correlations**: Confounding creates spurious correlations that machine learning models can mistake for genuine relationships.
> - **Feature Selection Errors**: Confounding can lead to the selection of irrelevant or even harmful features.
> - **Analogy: The Puppet Master**: A confounder is like a hidden puppet master, controlling both the treatment and the outcome, making it seem like the treatment causes the outcome when it doesn't.

3   The Backdoor Criterion

3.1   Definition

The backdoor criterion [1] provides a sufficient condition for identifying the causal effect of T on Y from observational data [cite: 3]. Given a causal directed acyclic graph (DAG), a set of variables Z satisfies the backdoor criterion relative to T and Y if:

1. No node in Z is a descendant of T.
2. Z blocks every path between T and Y that contains an arrow into T (i.e., every "backdoor path").

Blocking a path means that the path is not a channel for transmitting association. A path is blocked if it contains:

- A non-collider node that is conditioned on.
- A collider node that is not conditioned on, nor has any of its descendants conditioned on.

3.2 Backdoor Adjustment Formula

If Z satisfies the backdoor criterion, the causal effect of T on Y is given by the backdoor adjustment formula:

$$P(Y|do(T = t)) = \sum_z P(Y|T = t, Z = z)P(Z = z)$$

where $do(T = t)$ represents the intervention of setting T to the value t. This formula essentially averages the effect of T on Y within strata of Z, weighting by the distribution of Z.

3.3 Backdoor Adjustment with Regression Code Example

The backdoor adjustment can be implemented using regression. The coefficient of T in the regression of Y on T and Z estimates the causal effect of T on Y, provided Z satisfies the backdoor criterion.

Listing 7.1 Backdoor adjustment with regression

```python
import statsmodels.formula.api as sm  # For regression

# Simulate data
np.random.seed(10)
n = 1000
Z = np.random.normal(0, 1, n)  # Confounder
T = 0.5 * Z + np.random.normal(0, 1, n)
Y = 2 * T + Z + np.random.normal(0, 1, n)
data = pd.DataFrame({'Z': Z, 'T': T, 'Y': Y})

# 1. Naive Regression (biased)
model_naive = sm.ols("Y ~ T", data=data).fit()
print("Naive T coefficient:", model_naive.params['T'])

# 2. Regression Adjustment (unbiased, if Z blocks all
    backdoor paths)
model_adjusted = sm.ols("Y ~ T + Z", data=data).fit()
print("Adjusted T coefficient:", model_adjusted.params['T'])

# Explanation:
#    -    The coefficient of T in 'model_adjusted' is our
    estimate of the causal effect of T on Y,
#         *assuming* Z satisfies the backdoor criterion.
#    -    Statsmodels handles the summation implicitly through
     the regression.
```

> ### Causal and ML
>
> ### *Backdoor Criterion: The Recipe for Unbiasedness*
>
> - **Feature Selection Rule**: The backdoor criterion [1] provides a rule for selecting a set of features that, when included in a machine learning model, will yield unbiased estimates of the treatment effect.
> - **Model Design Principle**: It can also be seen as a principle for designing machine learning models, suggesting which inputs should be connected to which parts of the model.
> - **Analogy: The Filter**: The backdoor criterion is like a filter that removes the influence of confounding variables, allowing the model to focus on the true causal effect.

4   The Frontdoor Criterion

4.1   Definition

Sometimes, all backdoor paths from T to Y are blocked, but we can't measure the confounders. The frontdoor criterion [1] offers an alternative approach [cite: 3]. A set of variables M satisfies the frontdoor criterion relative to T and Y if:

1. All directed paths from T to Y pass through M.
2. There are no unblocked backdoor paths from T to M.
3. All backdoor paths from M to Y are blocked by T.

Here, M acts as a mediator, transmitting the causal effect of T on Y.

4.2   Frontdoor Adjustment Formula

If M satisfies the frontdoor criterion, the causal effect of T on Y can be estimated using the frontdoor adjustment formula:

$$P(Y|do(T)) = \sum_m P(M = m|do(T)) \sum_{t'} P(Y|do(T = t'), M = m) P(T = t')$$

This formula can be simplified under certain assumptions (e.g., linearity).

4.3 Code

Listing 7.2 Chapter 7: Focusing on demonstrating the backdoor and frontdoor criteria with simulated data and regression

```python
import numpy as np
import pandas as pd
import statsmodels.formula.api as sm  # For regression
import matplotlib.pyplot as plt
import seaborn as sns

# --- 1. Simulate Data for Backdoor Criterion ---
def simulate_backdoor_data(n=1000, effect_t_y=2, effect_z_t
    =0.5, effect_z_y=1):
    """Simulates data to demonstrate the backdoor criterion
        (Z -> T -> Y, Z -> Y)."""

    Z = np.random.normal(0, 1, n)  # Confounder
    T = 0.5 * Z + np.random.normal(0, 1, n)  # Treatment
        depends on Z
    Y = effect_t_y * T + effect_z_y * Z + np.random.normal
        (0, 1, n)
    return pd.DataFrame({'Z': Z, 'T': T, 'Y': Y})


# --- 2. Simulate Data for Frontdoor Criterion ---
def simulate_frontdoor_data(n=1000, effect_t_m=1.0,
    effect_m_y=1.5, effect_u_t=0.5, effect_u_y=0.8):
    """Simulates data to demonstrate the frontdoor criterion
        (T -> M -> Y, U -> T, U -> Y)."""

    U = np.random.normal(0, 1, n)  # Unobserved confounder
    T = effect_u_t * U + np.random.normal(0, 1, n)
    M = effect_t_m * T + np.random.normal(0, 1, n)  #
        Mediator
    Y = effect_m_y * M + effect_u_y * U + np.random.normal
        (0, 1, n)
    return pd.DataFrame({'U': U, 'T': T, 'M': M, 'Y': Y})


# --- 3. Demonstrate Backdoor Adjustment ---
def demonstrate_backdoor_adjustment(data):
    """Demonstrates backdoor adjustment using regression."""

    # 3.1. Naive regression (biased)
    model_naive = sm.ols('Y ~ T', data=data).fit()
    ate_naive = model_naive.params['T']
    print("\n--- Backdoor Criterion ---")
    print("Naive ATE (biased):", ate_naive)

    # 3.2. Adjusted regression (unbiased)
    model_adjusted = sm.ols('Y ~ T + Z', data=data).fit()
    ate_adjusted = model_adjusted.params['T']
```

```python
print("Adjusted ATE (unbiased):", ate_adjusted)

# 3.3. Visualization
plt.figure(figsize=(10, 5))

plt.subplot(1, 2, 1)
sns.regplot(x='T', y='Y', data=data, ci=95)
plt.title('Naive: Y vs. T')

plt.subplot(1, 2, 2)
sns.regplot(x='T', y='Y', data=data, ci=95)
sns.regplot(x='Z', y='Y', data=data, ci=None, color='red
    ')  # Show Z's influence
plt.title('Adjusted: Y vs. T and Z')

plt.tight_layout()
plt.show()

print("\nBackdoor adjustment corrects for confounding by
        including Z in the regression.")


# --- 4. Demonstrate Frontdoor Adjustment (Simplified) ---
def demonstrate_frontdoor_adjustment(data):
    """Demonstrates a simplified version of frontdoor
        adjustment using regression."""

    # 4.1. Estimate T -> M
    model_tm = sm.ols('M ~ T', data=data).fit()
    effect_t_to_m = model_tm.params['T']
    print("\n--- Frontdoor Criterion ---")
    print("Effect of T on M:", effect_t_to_m)

    # 4.2. Estimate M, T -> Y
    model_my = sm.ols('Y ~ M + T', data=data).fit()
    effect_m_to_y = model_my.params['M']
    print("Effect of M on Y (adjusted for T):",
        effect_m_to_y)

    # 4.3. Simplified frontdoor estimate (product of
        coefficients)
    frontdoor_estimate = effect_t_to_m * effect_m_to_y
    print("Simplified Frontdoor Estimate:",
        frontdoor_estimate)

    print("\nFrontdoor adjustment uses the mediator M to
        estimate the effect of T on Y, even with unobserved
        confounding U.")

    # 4.4. Visualization (Simplified)
    plt.figure(figsize=(8, 4))
    plt.subplot(1, 2, 1)
    sns.regplot(x='T', y='M', data=data, ci=95)
```

```python
    plt.title('T vs. M')

    plt.subplot(1, 2, 2)
    sns.regplot(x='M', y='Y', data=data, ci=95)
    sns.regplot(x='T', y='Y', data=data, ci=None, color='red
        ')  # Show T's direct influence
    plt.title('M vs. Y (adjusted for T)')

    plt.tight_layout()
    plt.show()


# --- 5. Main Execution ---
if __name__=="__main__":
    np.random.seed(123)

    # Backdoor Example
    backdoor_data = simulate_backdoor_data(effect_t_y=2,
        effect_z_t=0.5, effect_z_y=1)
    demonstrate_backdoor_adjustment(backdoor_data)

    # Frontdoor Example
    frontdoor_data = simulate_frontdoor_data(effect_t_m=1.0,
        effect_m_y=1.5, effect_u_t=0.5, effect_u_y=0.8)
    demonstrate_frontdoor_adjustment(frontdoor_data)
```

Causal and ML

Frontdoor Criterion: The Mediator Approach

These criteria provide a principled way to build machine learning models that are less susceptible to bias and more capable of generalization.

- **Mediator as Explanation**: The frontdoor criterion uses a mediator variable to understand how the treatment affects the outcome. This can lead to more interpretable machine learning models.
- **Step-by-Step Modeling**: It suggests a step-by-step approach to modeling, first modeling the effect of the treatment on the mediator, and then the effect of the mediator on the outcome.
- **Analogy: The Chain Reaction**: The frontdoor criterion is like tracing a chain reaction to understand how the treatment triggers the outcome.

5   Importance of Backdoor and Frontdoor Criteria

The backdoor and frontdoor criteria are essential tools for causal inference. They provide conditions under which causal effects can be identified from observational data. However, their correct application relies crucially on having a correct causal graph. Furthermore, the code examples provided are simplified illustrations. Real-world applications often involve more complex modeling techniques and careful consideration of the underlying assumptions.

6   Exercises

1. What is the challenge of confounding in causal inference, and why does it arise in observational data?
2. Define the backdoor criterion. What conditions must a set of variables satisfy to meet this criterion?
3. Explain the backdoor adjustment formula. How does it allow us to estimate causal effects in the presence of confounding?
4. Define the frontdoor criterion. What are the key differences between the frontdoor and backdoor criteria?
5. Under what conditions can the frontdoor criterion be used to estimate causal effects?

Reference

1. Pearl, Judea. 2009. *Causality: Models, reasoning, and inference*, 2nd. Cambridge University Press.

Advanced Causal Inference Methods

1 Instrumental Variables (IV)

1.1 Introduction and Context

Instrumental variables (IV) are statistical tools that help address confounding and identify causal effects in observational studies, especially when randomized trials are impractical or unethical. They work by isolating exogenous variation in the treatment variable, allowing researchers to estimate causal relationships reliably.

> **Causal and ML**
>
> ## IVs for Unobserved Confounding in ML
>
> - **The Hidden Bias**: Machine learning models can be severely biased by unobserved confounders, variables that influence both the input features and the outcome but are not included in the training data.
> - **IVs as a Solution**: Instrumental variables [1] provide a way to estimate causal effects even when such unobserved confounders are present, offering a more robust approach to causal inference in machine learning.
> - **Analogy: The Lever**: An instrumental variable is like a lever that we can use to move the treatment variable without directly affecting the outcome, allowing us to isolate the treatment's true causal effect.

D. Rajamanickam, *Causal Inference for Machine Learning Engineers*,
https://doi.org/10.1007/978-3-031-99680-1_8

1.2 Finding and Validating Instruments

Selecting an appropriate instrumental variable is crucial. An instrument must meet two critical criteria:

- **Relevance**: The instrument must affect the treatment. That is,

$$\mathrm{Cov}(Z, T) \neq 0$$

  where Z is the instrument and T is the treatment variable.
- **Exclusion Restriction**: The instrument affects the outcome only through its effect on the treatment, and not directly or through any other path. Formally,

$$Z \perp Y \mid T, U$$

  where Y is the outcome and U denotes unobserved confounders.

Real-World Example: Consider studying the effect of class size (T) on student test scores (Y). A potential confounder is student ability (U), which is unobserved and likely affects both class size (e.g., parents of high-ability students might advocate for smaller classes) and test scores. An instrumental variable could be a school construction lottery (Z), randomly assigning students to schools with different class sizes. The lottery outcome is correlated with class size (relevance). Still, it should not directly affect individual student test scores except through the class size to which they are assigned (exclusion restriction), assuming the lottery is fair (Fig. 1).

1.3 Weak Instrument Issues

Weak instruments, which are poorly correlated with the treatment, lead to biased and unreliable estimations. They can be identified by checking the F-statistic from the first-stage regression. An F-statistic below 10 typically indicates weak instrumentation.

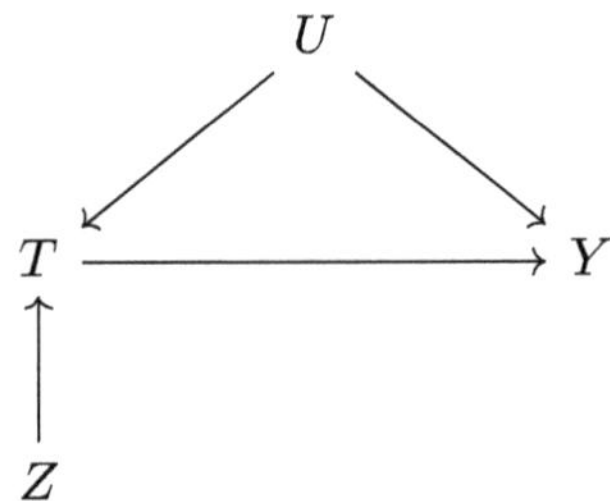

Fig. 1 Causal diagram: Lottery (Z) as an instrument for Class Size (T) affecting Test Scores (Y); U is an unobserved confounder

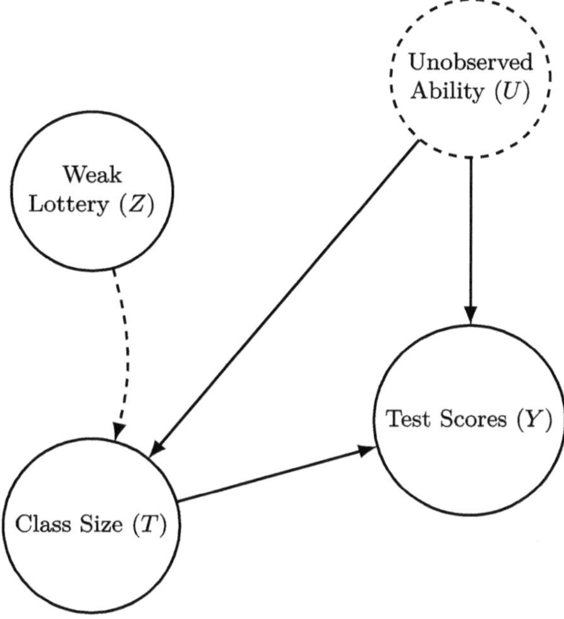

Fig. 2 Weak Instrument Scenario: the lottery instrument (Z) has a weak causal effect on class size (T)

> ## Causal and ML
>
> ### *IV Strength and ML Model Reliability*
>
> - **Weak Instruments = Noisy Information**: Weak instruments, which are poorly correlated with the treatment, lead to biased and unreliable estimations.
> - **Model Instability**: Using weak instruments in machine learning is analogous to training a model with highly noisy or irrelevant features, resulting in unstable and unreliable predictions.
> - **Analogy: The Unclear Signal**: A weak instrument is like trying to hear a faint whisper in a loud room; it's difficult to isolate the true message (the treatment effect).

Real-World Example: If, in the class size example, the school construction lottery (Z) had minimal impact on the actual class sizes students ended up in (e.g., due to many other factors influencing class assignment). The lottery would be a weak instrument. Using this weak instrument would likely produce unreliable estimates of the effect of class size on test scores, potentially even more biased than a simple OLS regression (Fig. 2).

2   Mediation Analysis

2.1   Formalization of Direct and Indirect Effects

Mediation analysis separates the total causal effect into direct and indirect, helping researchers understand how and why effects occur. Direct effects represent the effect of treatment on the outcome independent of the mediator, whereas indirect effects show the impact mediated through an intermediate variable.

2.2   Different Mediation Formulas

The key mediation formulas include:

$$TE = E[Y|do(T = 1)] - E[Y|do(T = 0)]$$
$$DE = E[Y|do(T = 1), M = m] - E[Y|do(T = 0), M = m]$$
$$IE = TE - DE$$

Real-World Example: Consider the effect of a job training program (T) on future earnings (Y). The program might directly improve skills leading to higher pay (direct effect). It might also increase job search effort (M) due to increased confidence, leading to better job opportunities and higher earnings (indirect effect mediated by job search effort) (Fig. 3).

2.3   Challenges in Mediation Analysis

Mediation analyses can be complicated by:

- Unmeasured confounding variables affecting mediator and outcome.
- Measurement error in mediator variables.
- Interaction effects between treatment and mediator.

Real-World Example (Unmeasured Confounding): In the job training example, an unmeasured factor like motivation could influence both participation in the training program and job search effort, as well as future earnings, potentially leading to spurious mediation effects if not accounted for (Fig. 4).

Fig. 3 Mediation diagram:
Job Training (T) affects
Future Earnings (Y) directly
and indirectly via Job Search
Effort (M)

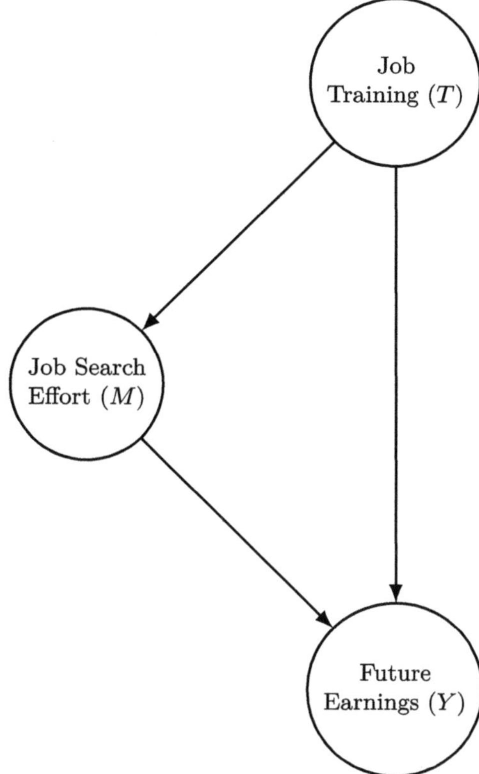

> ## Causal and ML
>
> ### *Mediation Analysis: Understanding Model Mechanisms*
>
> - **Causal Pathways in ML**: Mediation analysis helps us understand the
>   causal pathways within a machine learning model, identifying how input
>   features influence the outcome through intermediate representations or
>   calculations.
> - **Model Interpretability**: By decomposing the total effect into direct and
>   indirect effects, mediation analysis can improve the interpretability of com-
>   plex machine learning models, such as neural networks.
> - **Analogy: The Assembly Line**: Mediation analysis is like tracing the steps
>   in an assembly line to see how each stage contributes to the final product
>   (the model's prediction).

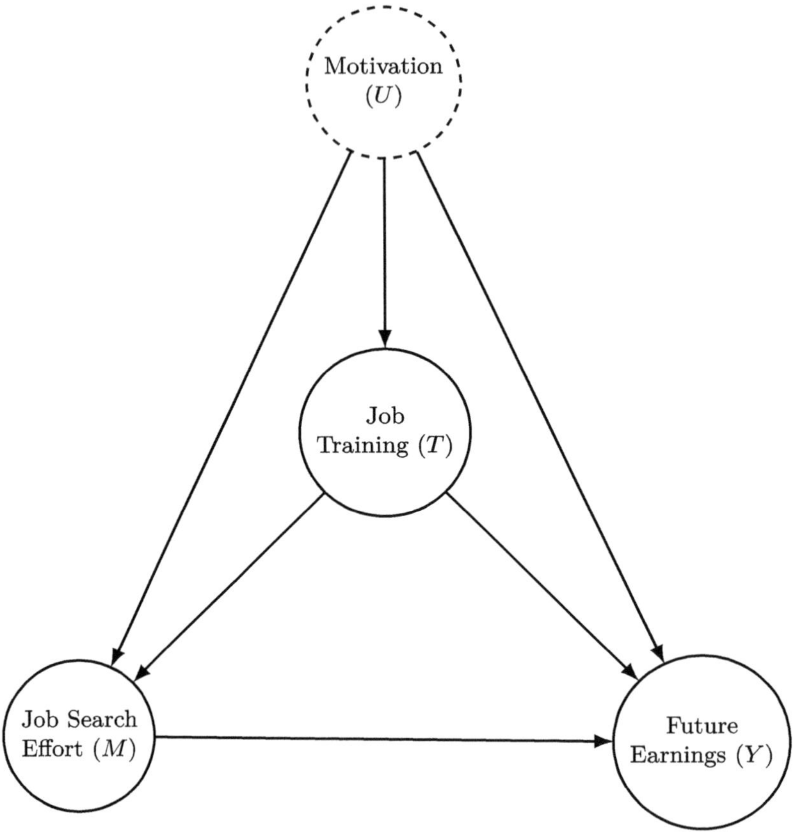

Fig. 4 Mediation with Unmeasured Confounding: Motivation (U) affects Job Training (T), Job Search Effort (M), and Future Earnings (Y)

3 Causal Discovery Algorithms

3.1 PC Algorithm

The PC constraint-based algorithm [2] identifies causal relationships by performing conditional independence tests among variables, constructing a Directed Acyclic Graph (DAG).

Real-World Example: Imagine researchers have data on air pollution levels (A), traffic volume (V), and respiratory illness rates (R) in a city. The PC algorithm [2] could be used to explore potential causal relationships. For instance, if A and R are correlated, but become independent when conditioning on V, it might suggest that traffic volume is a common cause of pollution and respiratory issues, or a mediator.

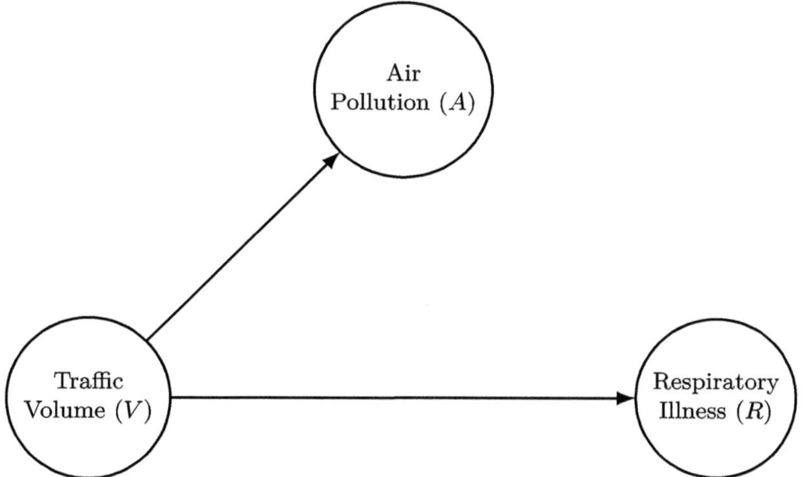

Fig. 5 Causal Discovery Example: Traffic Volume (V) affects both Air Pollution (A) and Respiratory Illness (R)

The PC algorithm would systematically test these conditional independencies to build a potential causal graph (Fig. 5).

3.2 FCI Algorithm

The Fast Causal Inference (FCI) algorithm extends the PC algorithm, accommodating scenarios where unobserved confounders and selection biases are present, thus providing a more comprehensive causal discovery [3] tool.

Real-World Example: Continuing the air pollution example, suppose there's an unmeasured factor (U) like industrial activity in different areas of the city that affects both traffic patterns (e.g., more trucks) and air pollution levels. The FCI algorithm, unlike PC, can potentially identify that the observed correlation between traffic and pollution might be partly due to this unobserved common cause, representing this uncertainty in its output graph (Partial Ancestral Graph) (Fig. 6).

3.3 LiNGAM

Linear Non-Gaussian Acyclic Models (LiNGAM) leverage non-Gaussian distribution properties to identify causal directions explicitly in linear models. This approach is robust when traditional methods fail due to issues with identifiability.

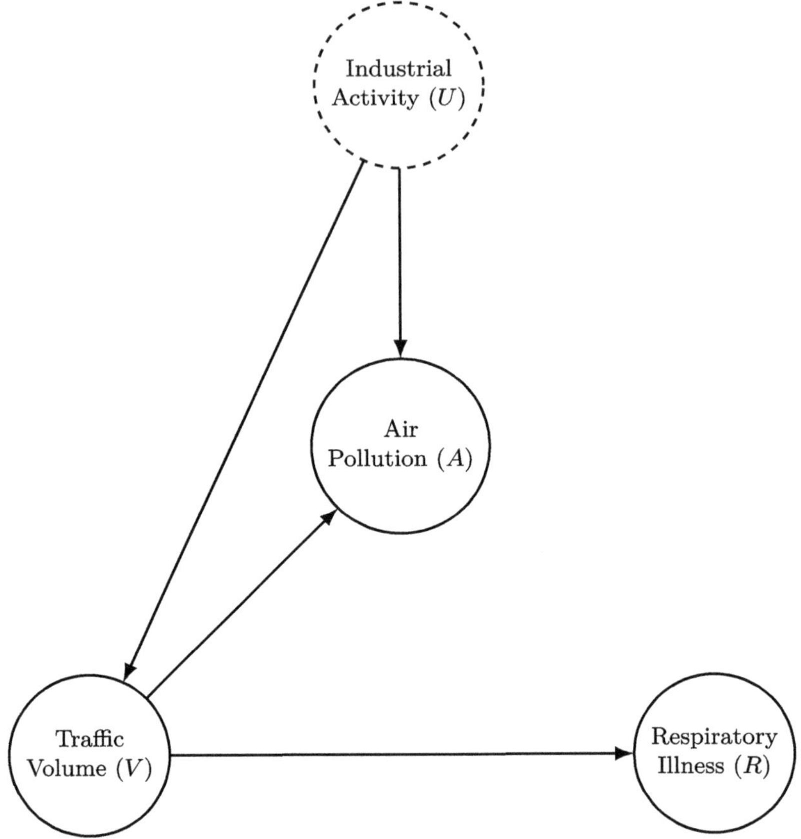

Fig. 6 Causal Discovery with a Latent Confounder: Industrial Activity (U) affects both Air Pollution (A) and Traffic Volume (V); V affects both A and Respiratory Illness (R)

Real-World Example: Consider the relationship between advertising expenditure (A) and product sales (S). Assuming a linear relationship and that the error terms affecting each are non-Gaussian (which is often plausible in economic data), LiNGAM could potentially determine if $A \rightarrow S$ (advertising drives sales) or $S \rightarrow A$ (higher sales allow for more advertising) or if there's a more complex bidirectional relationship (which LiNGAM, in its basic form, assumes is absent). The noise's non-Gaussianity helps break the symmetry inherent in purely correlational linear models with Gaussian noise (Fig. 7).

PC's output as a completed partially directed acyclic graph (CPDAG), not always a fully directed DAG.

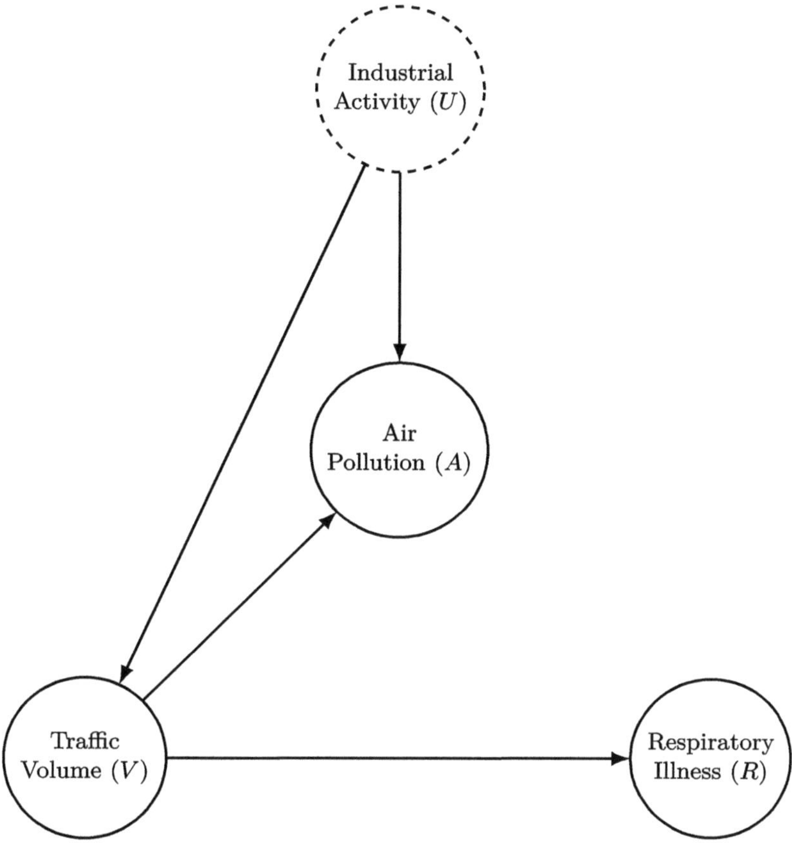

Fig. 7 Causal Discovery with a Latent Confounder: Industrial Activity (U) affects both Air Pollution (A) and Traffic Volume (V); V affects both A and Respiratory Illness (R)

Causal and ML

Causal Discovery: Learning Causal Structure from Data

- **Automated Causal Graph Construction**: Causal discovery [3]algorithms aim to automatically learn the causal graph from observational data, reducing the need for manual specification of causal relationships.
- **Feature Relationship Discovery**: These algorithms can help machine learning models discover the underlying causal relationships between input features, leading to more robust and generalizable representations.
- **Analogy: The Reverse Engineer**: Causal discovery [3]is like reverse-engineering a complex system to understand how its components interact.

4 Transportability

4.1 Generalizing Causal Knowledge

Transportability addresses the challenge of applying causal knowledge gained from one study or population to another distinct context. It adjusts for differences between source and target populations to ensure accurate causal inference.

4.2 Formalization

Transportability is formally represented as:

$$P_{target}(Y|do(T)) = \sum_{x_S} P(Y|do(T), X_S = x_S) P_{target}(X_S = x_S)$$

where the adjustment accounts for variables shared between the source and target populations. **Real-World Example**: Suppose a successful public health intervention (T) to increase vaccination rates (Y) was implemented in one city (source population). Researchers want to implement the same intervention in a different city (target population), but know that the demographic distribution of income levels (X_S) differs between the two cities. If income affects the effectiveness of the intervention or the baseline vaccination rates, transportability methods would use data from both cities on how income is related to vaccination and adjust the expected effect of the intervention in the target city based on its income distribution (Fig. 8).

Causal and ML

Transportability: Transferring Causal Knowledge

- **Causal Transfer Learning**: Transportability addresses the challenge of applying causal knowledge learned in one setting (source domain) to a different setting (target domain), which is analogous to transfer learning in machine learning.
- **Domain Adaptation**: This is particularly relevant in domain adaptation, where machine learning models must generalize to new environments with different data distributions.
- **Analogy: The Causal Travel Guide**: Transportability provides a "travel guide" that helps us adapt our causal knowledge to new "countries" (domains).

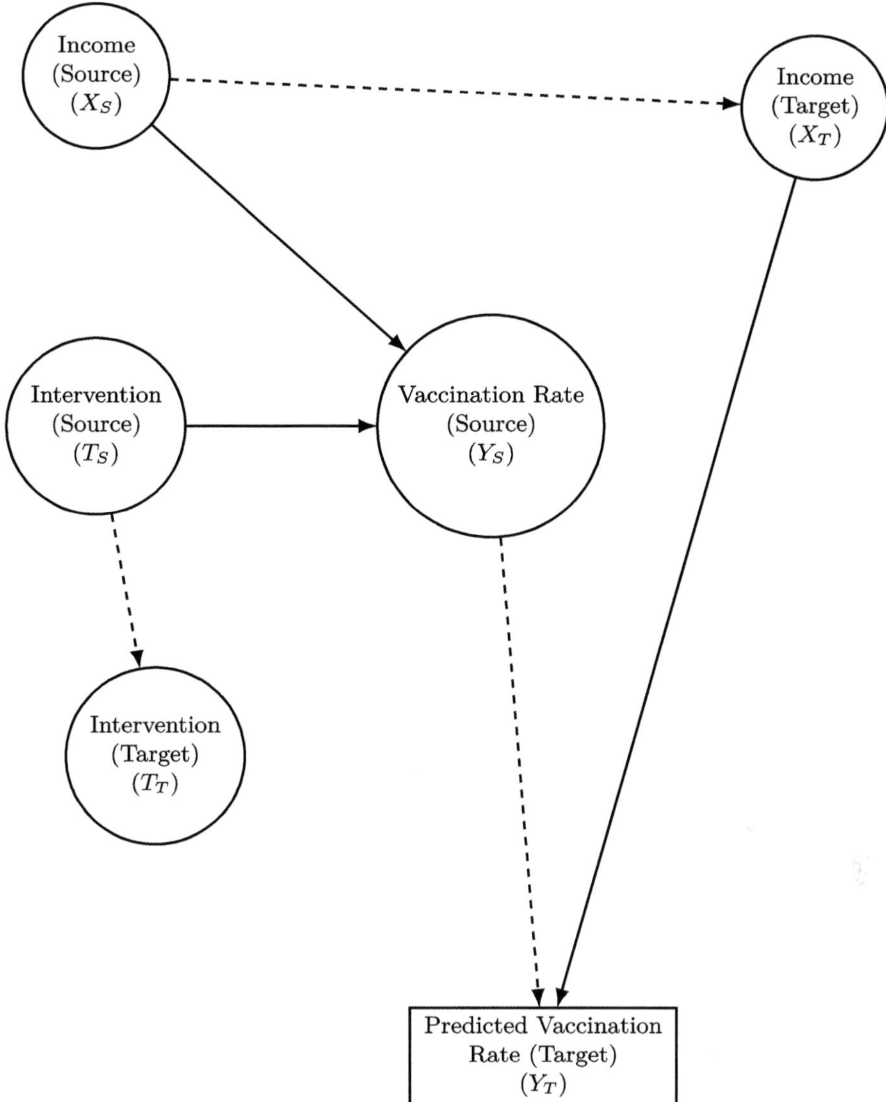

Fig. 8 Transportability Example: Predicting Target Vaccination Rate (Y_T) from Source Domain (T_S, X_S, Y_S) using shared covariates and structural similarities. Dashed arrows represent domain transport mappings

4.3 Code

Listing 8.1   Chapter 8: Covering Instrumental Variables (IV) and Mediation Analysis

```python
import numpy as np
import pandas as pd
import statsmodels.formula.api as sm
import matplotlib.pyplot as plt
import seaborn as sns

# --- 1. Simulate Data for Instrumental Variables ---
def simulate_iv_data(n=1000, effect_t_y=2, effect_z_t=0.5,
    effect_u_t=0.3, effect_u_y=0.8):
    """Simulates data for an Instrumental Variables scenario
        (Z -> T -> Y, U -> T, U -> Y).

    Args:
        n: Number of samples.
        effect_t_y: True causal effect of T on Y.
        effect_z_t: Effect of instrument Z on treatment T.
        effect_u_t: Effect of unobserved confounder U on T.
        effect_u_y: Effect of U on Y.

    Returns:
        Pandas DataFrame with Z, T, Y, and U.
    """

    U = np.random.normal(0, 1, n)   # Unobserved confounder
    Z = np.random.normal(0, 1, n)   # Instrument
    T = effect_z_t * Z + effect_u_t * U + np.random.normal
        (0, 1, n)
    Y = effect_t_y * T + effect_u_y * U + np.random.normal
        (0, 1, n)
    return pd.DataFrame({'Z': Z, 'T': T, 'Y': Y, 'U': U})


# --- 2. Implement Two-Stage Least Squares (2SLS) ---
def implement_2sls(data, plot_results=True):
    """Implements Two-Stage Least Squares (2SLS) to estimate
        the effect of T on Y.

    Args:
        data: DataFrame with Z, T, Y.
        plot_results: Whether to plot the stages.

    Returns:
        Estimated effect of T on Y.
    """

    # 2.1. First Stage: Regress T on Z
    first_stage = sm.ols('T ~ Z', data=data).fit()
    data['T_hat'] = first_stage.predict(data)  # Predicted T
```

```python
    # 2.2. Second Stage: Regress Y on T_hat
    second_stage = sm.ols('Y ~ T_hat', data=data).fit()
    effect_of_t = second_stage.params['T_hat']

    print("\n--- Two-Stage Least Squares (2SLS) ---")
    print("Estimated effect of T on Y:", effect_of_t)

    if plot_results:
        plt.figure(figsize=(12, 5))
        plt.subplot(1, 2, 1)
        sns.regplot(x='Z', y='T', data=data, ci=95)
        plt.title('First Stage: T vs. Z')
        plt.xlabel('Instrument (Z)')
        plt.ylabel('Treatment (T)')

        plt.subplot(1, 2, 2)
        sns.regplot(x='T_hat', y='Y', data=data, ci=95)
        plt.title('Second Stage: Y vs. Predicted T')
        plt.xlabel('Predicted Treatment (T_hat)')
        plt.ylabel('Outcome (Y)')

        plt.tight_layout()
        plt.show()

    return effect_of_t


# --- 3. Simulate Data for Mediation Analysis ---
def simulate_mediation_data(n=1000, effect_t_m=0.7,
    effect_t_y_direct=0.5, effect_m_y=0.8):
    """Simulates data for a mediation analysis scenario (T
        -> M -> Y, T -> Y).

    Args:
        n: Number of samples.
        effect_t_m: Effect of T on mediator M.
        effect_t_y_direct: Direct effect of T on Y.
        effect_m_y: Effect of M on Y.

    Returns:
        Pandas DataFrame with T, M, and Y.
    """

    T = np.random.normal(0, 1, n)
    M = effect_t_m * T + np.random.normal(0, 1, n)  #
        Mediator
    Y = effect_t_y_direct * T + effect_m_y * M + np.random.
        normal(0, 1, n)
    return pd.DataFrame({'T': T, 'M': M, 'Y': Y})


# --- 4. Implement Mediation Analysis (Linear Case) ---
def implement_mediation_analysis(data, plot_results=True):
```

```python
"""Implements a simplified mediation analysis (for
    linear case).

Args:
    data: DataFrame with T, M, Y.
    plot_results: Whether to plot the relationships.

Returns:
    Direct, indirect, and total effects.
"""

# 4.1. Estimate T -> M
model_tm = sm.ols('M ~ T', data=data).fit()
effect_t_to_m = model_tm.params['T']
print("\n--- Mediation Analysis ---")
print("Effect of T on M:", effect_t_to_m)

# 4.2. Estimate T, M -> Y
model_tmy = sm.ols('Y ~ T + M', data=data).fit()
effect_m_to_y = model_tmy.params['M']
effect_t_to_y_direct = model_tmy.params['T']
print("Effect of M on Y (adjusted for T):",
    effect_m_to_y)
print("Direct effect of T on Y:", effect_t_to_y_direct)

# 4.3. Calculate indirect effect
effect_t_to_y_indirect = effect_t_to_m * effect_m_to_y
effect_t_to_y_total = effect_t_to_y_direct + \
    effect_t_to_y_indirect
print("Indirect effect of T on Y (through M):",
    effect_t_to_y_indirect)
print("Total effect of T on Y:", effect_t_to_y_total)

if plot_results:
    plt.figure(figsize=(12, 4))
    plt.subplot(1, 3, 1)
    sns.regplot(x='T', y='M', data=data, ci=95)
    plt.title('T -> M')
    plt.xlabel('Treatment (T)')
    plt.ylabel('Mediator (M)')

    plt.subplot(1, 3, 2)
    sns.regplot(x='T', y='Y', data=data, ci=95)
    plt.title('Direct Effect of T on Y')
    plt.xlabel('Treatment (T)')
    plt.ylabel('Outcome (Y)')

    plt.subplot(1, 3, 3)
    sns.regplot(x='M', y='Y', data=data, ci=95)
    plt.title('M -> Y (adjusted for T)')
    plt.xlabel('Mediator (M)')
    plt.ylabel('Outcome (Y)')
```

```python
        plt.tight_layout()
        plt.show()

    return effect_t_to_y_direct, effect_t_to_y_indirect,
        effect_t_to_y_total


# --- 5. Main Execution ---
if __name__=="__main__":
    np.random.seed(42)  # Consistent seed

    # Instrumental Variables Example
    iv_data = simulate_iv_data(effect_t_y=2, effect_z_t=0.5,
        effect_u_t=0.3, effect_u_y=0.8)
    estimated_effect_iv = implement_2sls(iv_data)
    print("\nIn the IV example, Z is the instrument, U is
        the unobserved confounder.")
    print("2SLS attempts to recover the true effect of T on
        Y (which is 2 in this simulation).")

    # Mediation Analysis Example
    mediation_data = simulate_mediation_data(effect_t_m=0.7,
        effect_t_y_direct=0.5, effect_m_y=0.8)
    direct_effect, indirect_effect, total_effect =
        implement_mediation_analysis(mediation_data)
    print("\nIn the Mediation example, M is the mediator.")
    print("The code decomposes the total effect of T on Y
        into direct and indirect effects (through M).")
```

5 Difference-in-Differences (DiD)

5.1 *Motivation: Addressing Unobserved Confounding*

The Difference-in-Differences (DiD) method is a powerful quasi-experimental identification strategy, widely used when a standard Randomized Controlled Trial (RCT) is infeasible. DiD is designed to estimate the causal effect of a treatment or policy intervention (T) on an outcome (Y) using longitudinal or panel data where units are observed *before* and *after* the intervention.

The core challenge in observational causal inference is **unobserved confounding** (**U**). Standard methods like the Backdoor Criterion fail if a critical common cause is not measured. DiD addresses a specific type of unobserved confounding: that which is **time-invariant** (constant bias over time) and **group-invariant** (constant time trends across groups).

The fundamental idea of DiD is to create an appropriate **counterfactual trend** for the treatment group by observing the control group's trend over the same period.

5.2   The Core Methodology

The canonical DiD design requires a two-by-two structure:

1. **Two Groups:** A **Treatment Group** ($G = 1$) that receives the intervention, and a **Control Group** ($G = 0$) that does not.
2. **Two Periods:** A **Pre-intervention Period** ($t = 0$) and a **Post-intervention Period** ($t = 1$).

The treatment effect is estimated by calculating two successive differences, hence the name:

1. **First Difference (Time Trend):** This is the change in outcome over time for both the treatment and control groups individually.
2. **Second Difference (Group Comparison):** This is the difference between the two first differences. It isolates the change specific to the treatment group that did not occur in the control group.

The observed mean outcome $\bar{Y}_{G,t}$ for each cell of the 2×2 Table 1.

The DiD estimator $\hat{\tau}_{DiD}$ for the Average Treatment Effect on the Treated (ATT) is calculated as:

$$\hat{\tau}_{DiD} = (\bar{Y}_{T,Post} - \bar{Y}_{T,Pre}) - (\bar{Y}_{C,Post} - \bar{Y}_{C,Pre}) \tag{1}$$

This quantity represents the estimated causal effect of the intervention by removing the time-invariant selection bias (which causes the initial difference between groups) and the common time shock (which causes the change in the control group).

5.3   The Critical Parallel Trends Assumption

The causal validity of the DiD method rests entirely on the **Parallel Trends Assumption** (PTA).

5.3.1   Definition

The PTA states that, in the absence of the treatment, the unobserved potential outcomes for the treatment group would have followed the same time trend as the observed outcomes for the control group.

Table 1 The 2×2 DiD Setup

	Pre-Treatment ($t = 0$)	**Post-Treatment** ($t = 1$)
Control ($G = 0$)	$\bar{Y}_{C,Pre}$	$\bar{Y}_{C,Post}$
Treatment ($G = 1$)	$\bar{Y}_{T,Pre}$	$\bar{Y}_{T,Post}$

$$\mathbb{E}[Y_{T,Post}(0) - Y_{T,Pre}(0) \mid G = 1] = \mathbb{E}[Y_{C,Post}(0) - Y_{C,Pre}(0) \mid G = 0]$$

Where $Y(0)$ represents the potential outcome if no unit received the treatment. The term $\mathbb{E}[Y_{T,Post}(0) \mid G = 1]$ is the critical unobservable counterfactual that DiD seeks to identify using the observed trend of the control group.

5.3.2 Implications and Diagnostics

- **What it cancels:** The PTA effectively cancels out constant, unobserved differences (**U**) between the two groups (e.g., initial motivation levels, culture).
- **When it fails:** The assumption is violated if there is a unobserved factor whose influence on the outcome changes differently over time for the treatment group than for the control group. For instance, if the treatment group was already on a recovery trajectory independent of the intervention, the PTA is violated.
- **Pre-Trend Check (The Gold Standard):** The most important diagnostic is to look at multiple periods *before* the intervention. If the outcomes were trending in parallel during the pre-period, it makes the assumption that they would have continued to trend in parallel afterwards more credible.

5.4 DiD as a Regression Model

In practice, the DiD model is usually estimated via Ordinary Least Squares (OLS) regression, often referred to as a **Two-Way Fixed Effects (TWFE)** model when individual-level longitudinal data is available. This approach easily accommodates individual-level covariates ($\mathbf{X}_{it}$) and provides standard errors.

5.4.1 The Canonical Regression Equation

For a 2×2 design, the regression takes the form:

$$Y_{it} = \beta_0 + \beta_1 \cdot Treatment_i + \beta_2 \cdot Post_t + \delta \cdot (Treatment_i \times Post_t) + \mathbf{X}_{it}^{\mathsf{T}}\gamma + \epsilon_{it} \tag{2}$$

- β_0: Baseline outcome mean (Control Group, Pre-Period).
- β_1: Fixed difference between groups in the pre-period (Group-specific intercept).
- β_2: Fixed time trend common to both groups (Time-specific intercept).
- δ: **The DiD Estimand** (The Causal Effect). This coefficient measures the incremental change in the treatment group relative to the change in the control group.
- $\mathbf{X}_{it}^{\mathsf{T}}\gamma$: Inclusion of time-varying covariates $\mathbf{X}_{it}$ (optional, but standard practice for precision).

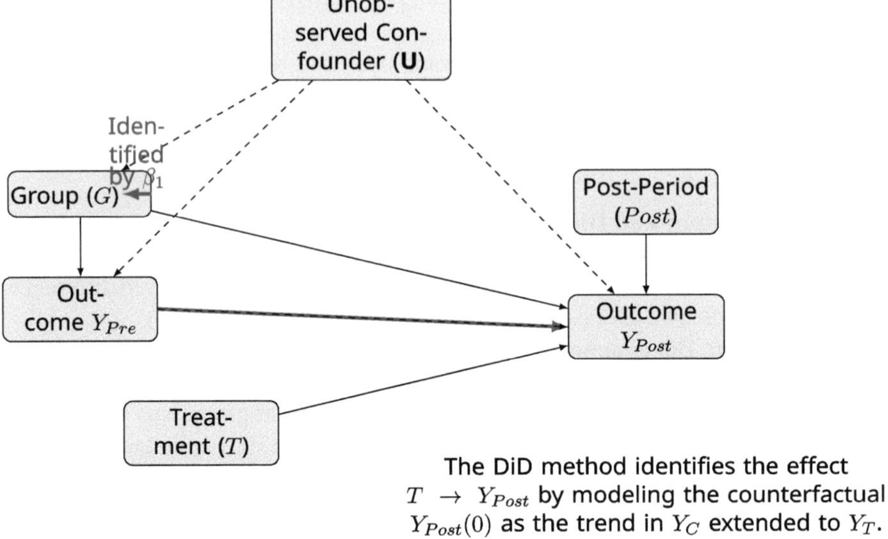

Fig. 9 Causal Diagram for a DiD Setup

5.4.2 Causal Diagram of the Setup

The DiD framework attempts to statistically implement the intervention concept by modeling the effect of T on Y while controlling for time-invariant unobserved confounding **U** (Fig. 9).

5.5 DiD and Machine Learning

While DiD is historically econometric, it holds strong relevance for modern causal machine learning:

- **Robustness to Unobserved Confounders:** DiD provides a structured way to handle unobserved confounders (a severe problem for many ML causal models) under the plausible condition that their influence is time-invariant.
- **Generalization to Staggered Adoption:** The base 2×2 model is extensible to modern panel data methods (like generalized linear models or causal forests) to handle situations where units adopt the treatment at different times (staggered DiD).
- **Causal Policy Evaluation:** DiD is the go-to tool for evaluating the causal impact of company-wide policy rollouts (e.g., changes to internal tools, new employee benefits) where unit-level randomization is not possible, providing a robust check for the estimated effect (δ).

6 Exercises

1. **Instrumental Variables**: Generate synthetic data violating IV assumptions [4] (e.g., instrument directly affects the outcome) and analyze how the violation impacts the 2SLS estimates. Illustrate the violation with a DAG.
2. **Mediation Analysis**: Conduct a mediation analysis with multiple mediators (e.g., job training $\rightarrow$ increased confidence $\rightarrow$ increased job search $\rightarrow$ better job $\rightarrow$ higher earnings).
3. **Causal Discovery**: Compare the performance of PC and FCI algorithms using simulated data with a known latent confounder.
4. **Transportability**: Simulate datasets reflecting two populations with different distributions of a key covariate. Apply the transportability formula to adjust causal effect estimates [5] and visualize the distributions and the adjustment process.

References

1. Angrist, Joshua D., G. W. Imbens, and D. B. Rubin. Identification of causal effects using instrumental variables. *Journal of the American Statistical Association* 91.434: 444–455.
2. Spirtes, Peter, and C. Glymour. 1991. An algorithm for fast recovery of sparse causal graphs. *Social Science Computer Review* 9.1: 62–72.
3. Spirtes, Peter, C. N. Glymour, and R. Scheines. 2000. *Causation, prediction, and search*. MIT Press.
4. Imbens, Guido W., and D. B. Rubin. 2015. *Causal inference for statistics, social, and biomedical sciences*. Cambridge University Press.
5. Pearl, Judea. 2009. *Causality: Models, Reasoning, and Inference*, 2nd. Cambridge University Press.

Causal Inference Meets Deep Learning

1 Introduction

In previous chapters, we explored the classical foundations of causal inference [1]: potential outcomes, interventions, and methods to estimate causal effects. These methods are powerful but are often designed for relatively simple, low-dimensional datasets where manual adjustment is feasible.

However, modern data, such as images, text, and sequential data, is increasingly high-dimensional and complex. In these settings, traditional causal methods face significant challenges. Meanwhile, deep learning has emerged as a transformative technology capable of learning rich, non-linear representations from raw data.

This chapter introduces how deep learning can be integrated with causal inference [1], the challenges involved, and the specialized models developed to bridge these two fields.

2 Why Deep Learning?

Deep learning [2] refers to machine learning models based on neural networks with many layers, capable of automatically learning useful features from raw data. In standard machine learning, the goal is to predict outcomes accurately. However, causal inference aims to estimate counterfactual outcomes—what would have happened under alternative treatments.

Deep learning offers two primary advantages for causal inference:

- **Representation Learning**: Deep networks can extract high-level features from complex data without manual engineering.
- **Flexibility**: They can model highly non-linear relationships between covariates, treatments, and outcomes.

D. Rajamanickam, *Causal Inference for Machine Learning Engineers*,
https://doi.org/10.1007/978-3-031-99680-1_9

Nevertheless, the fundamental challenge remains: while deep models are excellent predictors, they must be carefully adapted to estimate causal effects, not merely fit observational outcomes.

The ability of deep learning to learn complex representations is beneficial for causal inference in high-dimensional and unstructured data, where traditional methods struggle.

Deep Learning for Causal Tasks

- **Representation Learning Power**: Deep learning's ability to automatically learn useful feature representations (a core concept in deep learning) is crucial for handling complex data in causal inference.
- **Beyond Prediction**: While standard deep learning focuses on prediction, causal deep learning adapts these models to estimate counterfactuals and treatment effects, going beyond pattern recognition.

3   Challenges in Causal Deep Learning

Applying deep learning to causal inference introduces several challenges:

- **Selection Bias**: Observational data often exhibit bias in treatment assignment. Without correction, deep models replicate this bias.
- **Counterfactual Estimation**: Only one potential outcome is observed for each individual; the missing counterfactual must be inferred.
- **Overfitting and Generalization**: Deep models with high capacity risk overfitting observed data, can worsen counterfactual predictions.

These challenges motivate the development of specialized causal deep learning models [2].

Causal deep learning aims to address the limitations of standard deep learning models, such as their susceptibility to bias and lack of interpretability.

4   Treatment-Agnostic Representation Networks (TARNet)

One of the earliest models addressing causal inference with deep learning is the **Treatment-Agnostic Representation Network (TARNet)** is a deep learning architecture that learns a shared representation of covariates, followed by treatment-specific outcome [3].

TARNet architecture consists of:

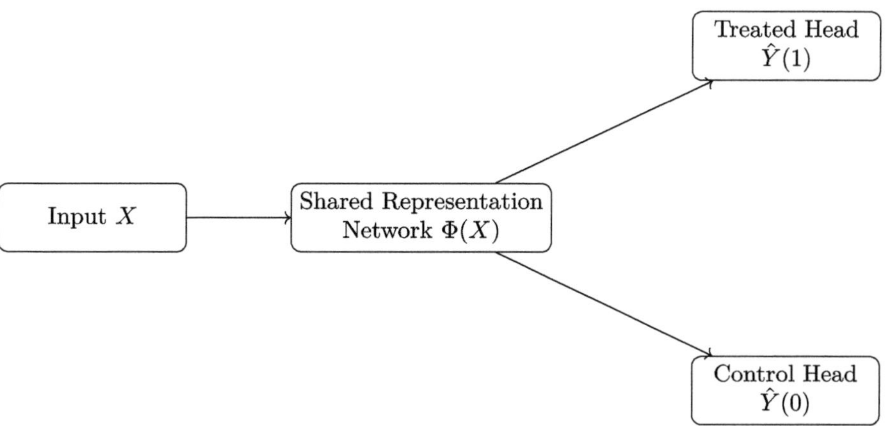

Fig. 1 TARNet Architecture: a shared representation network followed by two treatment-specific heads

- A shared feature extractor $\Phi(X)$ that learns a representation of covariates X.
- Two output heads:

$$\hat{Y}(0) = f_0(\Phi(X)), \quad \hat{Y}(1) = f_1(\Phi(X))$$

where f_0 and f_1 predict outcomes under control and treatment, respectively.

As shown in Fig. 1, TARNet aims to learn a representation helpful in predicting outcomes under both treatment and control. However, it does not explicitly enforce balance between the treated and control groups.

TARNet: Adapting Representation Learning

- **Shared Representations**: TARNet uses a shared representation network (a common architecture in deep learning) to learn relevant features for both treated and control outcomes.
- **Treatment-Specific Heads**: The two output heads are similar to having separate regressors in machine learning, but they share the learned representation.

5 Counterfactual Regression Networks (CFRNet)

To address covariate imbalance, **Counterfactual Regression Networks (CFRNet)** extend TARNet by introducing a distributional regularization term.

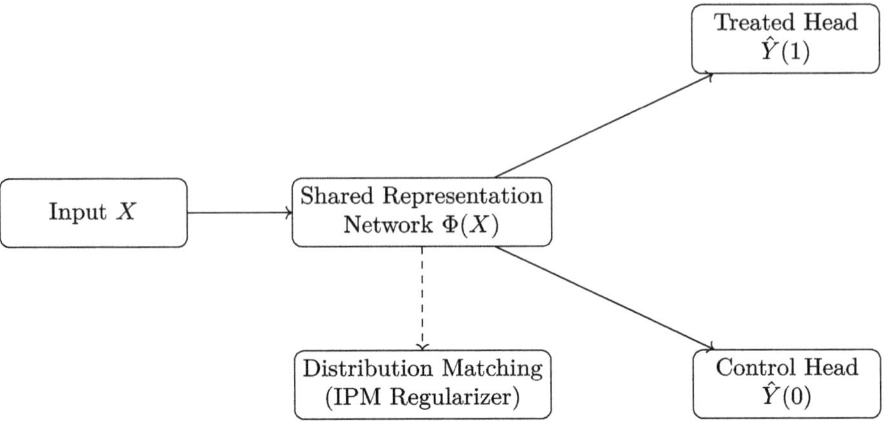

Fig. 2 CFRNet Architecture: TARNet plus distribution matching using an Integral Probability Metric (IPM) to balance representations

The CFRNet [3] loss function is:

$$\mathcal{L}_{\text{CFRNet}} = \mathcal{L}_{\text{prediction}} + \alpha \cdot \text{IPM}(p(\Phi(X) \mid T = 1), p(\Phi(X) \mid T = 0)) \qquad (1)$$

where:

- $\mathcal{L}_{\text{prediction}}$ is the outcome prediction loss.
- IPM is an Integral Probability Metric (e.g., Maximum Mean Discrepancy or Wasserstein distance) measuring the difference between representations.
- α controls the strength of the regularization.

Figure 2 illustrates CFRNet, where representations are explicitly regularized to encourage balance between treatment groups.

CFRNet: Balancing with Metrics

- **Distributional Regularization**: CFRNet adds a regularization term, similar to weight decay in deep learning. However, this term explicitly encourages the representations of treated and control groups to be similar, using Integral Probability Metrics (IPMs).
- **Adversarial Connection**: This balancing act has connections to adversarial training in deep learning, where models are trained to be robust to perturbations.

6 DragonNet: Learning Propensity Scores

DragonNet [4] enhances causal estimation by jointly learning potential outcomes and the propensity score $e(X) = \mathbb{P}(T = 1 \mid X)$.

The DragonNet [4] architecture includes:

- A shared representation network.
- Two outcome heads for treated and control outcomes.
- An additional propensity score head.

The loss function combines:

- Outcome prediction loss.
- Binary cross-entropy loss for propensity score estimation.
- Targeted regularization to improve counterfactual predictions.

As shown in Fig. 3, DragonNet explicitly models the treatment assignment mechanism, improving robustness to confounding.

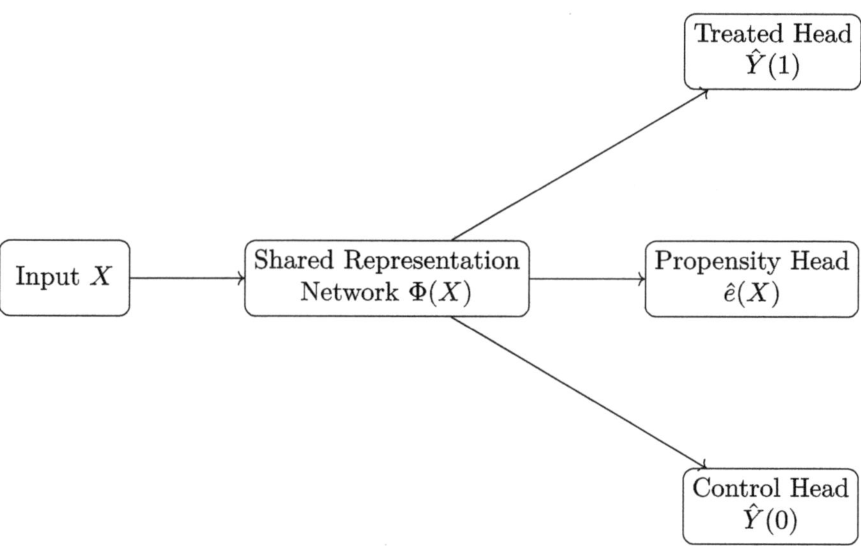

Fig. 3 DragonNet Architecture: outcome prediction and propensity score estimation with targeted regularization

DragonNet: Multi-Task Learning for Causality

- **Multi-Task Architecture**: DragonNet is a multi-task learning model (a deep learning technique) that simultaneously predicts outcomes and propensity scores, improving confounding adjustment.
- **Propensity as Regularizer**: The propensity score prediction acts as a regularizer, guiding the representation learning to capture treatment assignment.

7 Double Machine Learning (DoubleML)

Double Machine Learning (DoubleML) [5] provides a flexible, two-stage framework for causal estimation [5].

The key steps in DoubleML are:

1. Estimate nuisance parameters:

$$\hat{e}(X) \quad \text{(propensity score)}, \quad \hat{m}(X) \quad \text{(outcome model)}$$

using flexible machine learning models.
2. Compute residuals:

$$\tilde{T} = T - \hat{e}(X), \quad \tilde{Y} = Y - \hat{m}(X)$$

3. Regress $\tilde{Y}$ on $\tilde{T}$ to estimate the causal effect.

The orthogonalization [5] step ensures that minor errors in estimating $\hat{e}(X)$ or $\hat{m}(X)$ do not bias the final causal effect estimator.

Table 1 summarizes the key causal deep learning methods discussed in this chapter.

Table 1 Comparison of methods for causal deep learning

Method	Main idea	Advantages
TARNet	Shared representation, two outcome heads	Simple, flexible
CFRNet	TARNet + balanced representations	Reduces confounding bias
DragonNet	Outcome + Propensity score prediction	Better adjustment for confounding
DoubleML	Orthogonalized ML estimation	Robust to model misspecification

8 Double Machine Learning for Causal Inference

8.1 Motivation

When estimating causal effects with high-dimensional covariates, naive machine learning models tend to introduce bias. **Double Machine Learning (DML)** [5] was developed to correct for this bias and enable valid causal estimation even when using flexible, high-capacity models like random forests or neural networks [5].

DML separates prediction errors from causal estimation errors by applying the principle of **Neyman Orthogonality**: Small errors in nuisance models (outcome model, propensity model) do not bias the treatment effect estimate.

DoubleML: Debiased Learning

- **Orthogonalization**: DoubleML employs orthogonalization to de-bias the causal effect estimates, making them robust to errors in the nuisance function estimation (propensity score and outcome model).
- **Meta-Algorithm**: DoubleML is more of a meta-algorithm that can be used with various machine learning models (e.g., neural networks, random forests) for causal inference.

8.2 Key Idea

The core steps of DML are:

1. Predict the outcome Y given covariates X using any ML model: $\hat{g}(X)$.
2. Predict the treatment T given covariates X using any ML model: $\hat{e}(X)$ (propensity score model).
3. Form residuals:
$$\tilde{Y} = Y - \hat{g}(X), \quad \tilde{T} = T - \hat{e}(X)$$

4. Estimate the treatment effect θ by regressing $\tilde{Y}$ on $\tilde{T}$.

This two-stage procedure ensures that the treatment effect estimator is **robust to minor errors** in nuisance parameter estimation.

8.3 DML Pipeline Diagram

See Fig. 4.

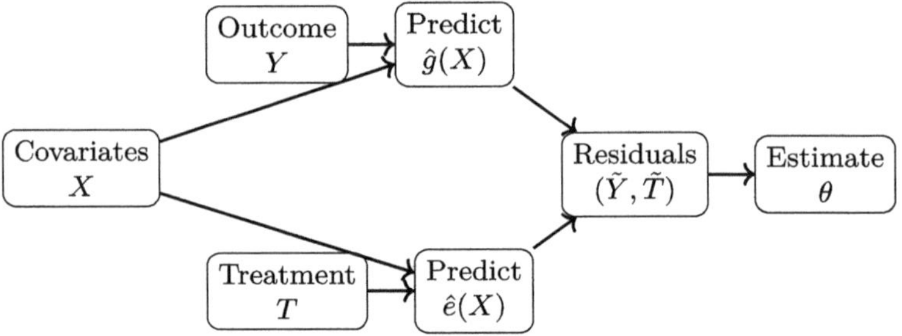

Fig. 4 Double Machine Learning pipeline: nuisance estimation followed by residual regression

8.4 Double ML: Code Example

Listing 9.1 Simple double machine learning implementation

```python
import numpy as np
from sklearn.linear_model import LinearRegression
from sklearn.ensemble import RandomForestRegressor
from sklearn.model_selection import train_test_split

# Simulate data
np.random.seed(42)
n = 1000
X = np.random.normal(0, 1, size=(n, 5))
T = np.random.binomial(1, 0.5, size=n)
Y = 2 * T + np.dot(X, [0.5, -0.3, 0.2, 0.1, -0.2]) + np.
    random.normal(0, 1, n)

# Step 1: Fit nuisance models
g_model = RandomForestRegressor().fit(X, Y)
e_model = RandomForestRegressor().fit(X, T)

# Step 2: Compute residuals
g_pred = g_model.predict(X)
e_pred = e_model.predict(X)

Y_res = Y - g_pred
T_res = T - e_pred

# Step 3: Regress residuals
reg = LinearRegression().fit(T_res.reshape(-1, 1), Y_res)
theta_hat = reg.coef_[0]

print(f"Estimated treatment effect (theta): {theta_hat:.4f}"
    )
```

8.5 Relationship to Deep Learning Methods

While methods like CFRNet [6] and DragonNet [4] aim to balance representations across treatment groups, **Double Machine Learning** [5] directly de-biases the estimation procedure through orthogonalization [5].

Both strategies aim to solve the underlying problem: obtaining valid causal estimates in complex, high-dimensional data.

9 Summary

This chapter introduced the interface between causal inference and deep learning. We discussed how deep learning models can help causal estimation through flexible representation learning, but also face challenges such as bias and counterfactual estimation.

We introduced TARNet [3], CFRNet] [6], and DragonNet [4] as architectures tailored for causal tasks, and discussed DoubleML [5] as a robust estimation framework.

In the next chapters, we will dive deeper into simulating causal datasets and evaluating causal models in practice.

Causal deep learning holds the potential to create more robust, reliable, and trustworthy machine learning systems for a wide range of applications.

10 Code

Listing 9.2 Chapter 7: a basic TARNet-like structure

```
import torch
import torch.nn as nn
import torch.optim as optim
from torch.utils.data import DataLoader, TensorDataset

import numpy as np
import pandas as pd
import matplotlib.pyplot as plt
from sklearn.model_selection import train_test_split
from sklearn.metrics import mean_squared_error

# --- 1. Simulate Data ---
def simulate_treatment_effect_data(n=1000, input_dim=5, ate
    =2.0):
    """Simulates data with treatment effect, similar to IHDP

    .

    Args:
```

```python
        n: Number of samples.
        input_dim: Number of input features.
        ate: Average Treatment Effect.

    Returns:
        X, T, Y, Y0, Y1 (torch tensors)
    """

    X = np.random.rand(n, input_dim).astype(np.float32)  #
        Covariates
    T = np.random.binomial(1, 0.5, n).astype(np.float32)  #
        Treatment (binary)
    Y0 = np.dot(X, np.random.rand(input_dim, 1)).flatten() +
        np.random.normal(0, 1, n)  # Outcome if T=0
    Y1 = Y0 + ate + np.random.normal(0, 1, n)  # Outcome if
        T=1
    Y = T * Y1 + (1 - T) * Y0  # Observed outcome

    # Explicitly specify dtype=torch.float32 for all tensors
    X_tensor = torch.tensor(X, dtype=torch.float32)
    T_tensor = torch.tensor(T, dtype=torch.float32).reshape
        (-1, 1)
    Y_tensor = torch.tensor(Y, dtype=torch.float32).reshape
        (-1, 1)
    Y0_tensor = torch.tensor(Y0, dtype=torch.float32).
        reshape(-1, 1)  # For evaluation
    Y1_tensor = torch.tensor(Y1, dtype=torch.float32).
        reshape(-1, 1)

    return X_tensor, T_tensor, Y_tensor, Y0_tensor,
        Y1_tensor

# --- 2. TARNet Architecture ---
class TARNet(nn.Module):
    def __init__(self, input_dim, hidden_dim=100):
        super(TARNet, self).__init__()
        self.shared_representation = nn.Sequential(
            nn.Linear(input_dim, hidden_dim),
            nn.ReLU(),
            nn.Linear(hidden_dim, hidden_dim),
            nn.ReLU()
        )
        self.outcome_T0 = nn.Linear(hidden_dim, 1)  #
            Outcome head for T=0
        self.outcome_T1 = nn.Linear(hidden_dim, 1)  #
            Outcome head for T=1

    def forward(self, x, t):
        shared = self.shared_representation(x)
        y0_hat = self.outcome_T0(shared)
        y1_hat = self.outcome_T1(shared)
        y_hat = t * y1_hat + (1 - t) * y0_hat  # Observed
            outcome prediction
```

```python
        return y_hat, y0_hat, y1_hat


# --- 3. CFRNet Architecture (Simplified Balancing) ---
class CFRNet(nn.Module):
    def __init__(self, input_dim, hidden_dim=100, ipm_lambda
        =0.1):  # ipm_lambda: balancing strength
        super(CFRNet, self).__init__()
        self.shared_representation = nn.Sequential(
            nn.Linear(input_dim, hidden_dim),
            nn.ReLU(),
            nn.Linear(hidden_dim, hidden_dim),
            nn.ReLU()
        )
        self.outcome_T0 = nn.Linear(hidden_dim, 1)
        self.outcome_T1 = nn.Linear(hidden_dim, 1)
        self.ipm_lambda = ipm_lambda

    def forward(self, x, t):
        shared = self.shared_representation(x)
        y0_hat = self.outcome_T0(shared)
        y1_hat = self.outcome_T1(shared)
        y_hat = t * y1_hat + (1 - t) * y0_hat

        return y_hat, y0_hat, y1_hat, shared  # Return
            shared for balancing

    def calculate_balancing_loss(self, shared, t):
        """Simplified balancing loss: difference in means.
        """
        shared_t1 = shared[t.flatten() == 1]
        shared_t0 = shared[t.flatten() == 0]
        if shared_t1.shape[0] == 0 or shared_t0.shape[0] ==
            0:  # Handle edge case
            # Ensure the returned tensor also has the
                correct dtype
            return torch.tensor(0.0, requires_grad=True,
                dtype=torch.float32)
        # Ensure the result of the mean difference is also
            float32
        return (torch.mean(shared_t1, dim=0) - torch.mean(
            shared_t0, dim=0))

# --- 4. Training Function ---
def train_model(model, X_train, T_train, Y_train, optimizer,
    epochs=100, batch_size=32, model_type="TARNet"):
    """Trains TARNet or CFRNet."""

    dataset = TensorDataset(X_train, T_train, Y_train)
    dataloader = DataLoader(dataset, batch_size=batch_size,
        shuffle=True)
    criterion = nn.MSELoss()
    model.train()
```

```python
    for epoch in range(epochs):
        for x_batch, t_batch, y_batch in dataloader:
            optimizer.zero_grad()
            if model_type == "CFRNet":
                y_pred, _, _, shared_batch = model(x_batch,
                    t_batch)
                prediction_loss = criterion(y_pred, y_batch)
                # Ensure balancing_loss is computed and
                    added as float32
                balancing_loss = model.
                    calculate_balancing_loss(shared_batch,
                    t_batch).mean()
                loss = prediction_loss + model.ipm_lambda *
                    balancing_loss
            else:  # TARNet
                y_pred, _, _ = model(x_batch, t_batch)
                loss = criterion(y_pred, y_batch)

            loss.backward()
            optimizer.step()
        if (epoch + 1) % 10 == 0:
            print(f'Epoch {epoch + 1}, Loss: {loss.item()}')


# --- 5. Evaluation Function ---
def evaluate_model(model, X_test, T_test, Y_test, Y0_test,
    Y1_test):
    """Evaluates the trained model."""

    model.eval()
    with torch.no_grad():
        # Check model type to handle different return values
        if isinstance(model, CFRNet):
            y_pred, y0_pred, y1_pred, _ = model(X_test,
                T_test) # Unpack 4, ignore the 4th
        else: # Assume TARNet or similar returning 3 values
            y_pred, y0_pred, y1_pred = model(X_test, T_test)

        y_pred = y_pred.squeeze().numpy()
        y0_pred = y0_pred.squeeze().numpy()
        y1_pred = y1_pred.squeeze().numpy()
        Y_test = Y_test.squeeze().numpy()
        Y0_test = Y0_test.squeeze().numpy()
        Y1_test = Y1_test.squeeze().numpy()

        mse = mean_squared_error(Y_test, y_pred)
        ate_pred = np.mean(y1_pred - y0_pred)
        ate_true = np.mean(Y1_test - Y0_test)
        ate_error = np.abs(ate_pred - ate_true)

        print("\n--- Evaluation ---")
        print(f"MSE: {mse:.4f}")
```

```python
        print(f"Predicted ATE: {ate_pred:.4f}")
        print(f"True ATE: {ate_true:.4f}")
        print(f"ATE Error: {ate_error:.4f}")

# --- 6. Main Execution ---
if __name__=="__main__":
    np.random.seed(42)
    X, T, Y, Y0, Y1 = simulate_treatment_effect_data()
    X_train, X_test, T_train, T_test, Y_train, Y_test,
        Y0_train, Y0_test, Y1_train, Y1_test =
        train_test_split(
        X, T, Y, Y0, Y1, test_size=0.2
    )

    # 6.1. Train and Evaluate TARNet
    tarnet = TARNet(input_dim=X.shape[1])
    optimizer_tarnet = optim.Adam(tarnet.parameters(), lr
        =0.01)
    train_model(tarnet, X_train, T_train, Y_train,
        optimizer_tarnet, model_type="TARNet")
    evaluate_model(tarnet, X_test, T_test, Y_test, Y0_test,
        Y1_test)

    # 6.2. Train and Evaluate CFRNet
    cfrnet = CFRNet(input_dim=X.shape[1], ipm_lambda=0.1)   #
        Experiment with ipm_lambda
    optimizer_cfrnet = optim.Adam(cfrnet.parameters(), lr
        =0.01)
    train_model(cfrnet, X_train, T_train, Y_train,
        optimizer_cfrnet, model_type="CFRNet")
    # Correcting the order of arguments in evaluate_model
        for CFRNet
    evaluate_model(cfrnet, X_test, T_test, Y_test, Y0_test,
        Y1_test)

    print("\n--- Explanation ---")
    print("This code demonstrates simplified TARNet and
        CFRNet implementations.")
    print("TARNet learns a shared representation and has
        separate outcome heads.")
    print("CFRNet adds a balancing loss to TARNet to
        encourage similar representations for treated and
        control groups.")
    print("The simulation generates synthetic data with a
        known Average Treatment Effect (ATE).")
    print("Evaluation compares the models' ability to
        predict outcomes and estimate the ATE.")
```

11 Exercises

1. Why is deep learning useful for causal inference in high-dimensional settings?
2. Explain the key challenges in applying deep learning to causal problems.
3. Describe the purpose of representation learning in causal models.
4. What could go wrong if deep learning models ignore treatment assignment bias?

References

1. Pearl, Judea. 2009. *Causality: Models, reasoning, and inference*, 2nd. Cambridge University Press.
2. Goodfellow, Ian, Y. Bengio, and A. Courville. 2016. *Deep learning*. MIT Press.
3. Shalit, Uri, F. D. Johansson, and D. Sontag. 2017. Estimating individual treatment effect: Generalization bounds and algorithms . In: *Proceedings of the 34th international conference on machine learning (ICML)*, 3076–3085.
4. Shi, Chao, D. M. Blei, and V. Veitch. 2019. Adapting neural networks for the estimation of treatment effects . In: *Advances in neural information processing systems (NeurIPS)*.
5. Chernozhukov, Victor et al. 2018. Double/debiased machine learning for treatment and structural parameters . *The Econometrics Journal* 21.1: C1 C68.
6. Johansson, Fredrik D., U. Shalit, and D. Sontag. 2016. Learning representations for counterfactual inference . In: *Proceedings of the 33rd international conference on machine learning (ICML)*, 1386–1395.

Simulating Causal Data and Evaluation Metrics

1 Introduction

The core challenge in causal inference [1] is the problem of missing counterfactuals. In any real-world dataset, for each individual, we observe only the outcome under the treatment they received. We cannot observe what would have happened had the individual received a different treatment. This problem is sometimes called **The Fundamental Problem of Causal Inference**.

Because of this, it becomes tough to evaluate how well a causal model is performing. How can we check whether our model's estimates are correct if we do not know the true causal effects?

A critical solution to this dilemma is to create **simulated datasets**, where we, the data creators, control the entire data-generating process. In simulated data, we know both the factual and counterfactual outcomes because we define them. This enables us to measure how close a model's estimates are to the true causal effects.

This chapter will teach us how to simulate causal data, understand the basic ideas behind synthetic benchmarks, and introduce the key metrics used to evaluate causal inference [1] models.

2 Why Simulate Causal Data?

Before building simulation procedures, it is helpful to understand why simulation is so valuable in causal inference [1].

In traditional supervised learning tasks like image classification, the true label for each input (e.g., whether an image contains a cat or not) is known. We can easily compute a model's accuracy by comparing its predictions to the true labels.

D. Rajamanickam, *Causal Inference for Machine Learning Engineers*,
https://doi.org/10.1007/978-3-031-99680-1_10

In causal inference [1], however, the true labels—the counterfactual outcomes—are *unobservable*. We can only see the outcome corresponding to the treatment that was assigned.

Simulating data allows us to bypass this limitation by constructing artificial worlds where the ground truth causal effects are fully known. In these synthetic settings:

- We can generate both the potential outcomes: $Y(0)$ and $Y(1)$.
- We can assign treatments according to a known (possibly biased) mechanism.
- We can add controlled amounts of noise or confounding to study their effects.

Thus, simulation is a laboratory for developing, testing, and validating causal inference methods under controlled conditions.

Simulation for Causal ML Evaluation

- **The Challenge of Unobservables**: In standard supervised learning, we have ground truth labels. In causal inference, counterfactuals are unobserved, making direct evaluation tricky. Simulation provides a "sandbox" with known ground truth.
- **Controlling the Data Pipeline**: Simulation lets us control every aspect of the data generation, similar to how we create synthetic data for training and testing machine learning models in specific scenarios (e.g., image augmentation).
- **Robustness Testing**: We can simulate various levels of confounding or noise to test the robustness of causal machine learning models, analogous to testing a classifier's robustness to adversarial attacks.

3 Building a Simple Simulation

Let us now walk through a basic example of simulating causal data step by step.

First, we define a set of covariates, which are features describing individuals. For simplicity, suppose each individual's features X are drawn from a standard normal distribution:

$$X \sim \mathcal{N}(0, I_d)$$

where d is the number of covariates, and I_d is the identity matrix, the features are independent and each has unit variance.

Next, we define how treatment is assigned. In real-world data, treatment assignment is rarely random. It often depends on an individual's features. To mimic this, we define a treatment assignment probability using a logistic function:

$$\mathbb{P}(T = 1 \mid X) = \mathrm{sigmoid}(w^\top X)$$

Fig. 1 Simulated data-generating process: Covariates X influence both treatment T and outcome Y

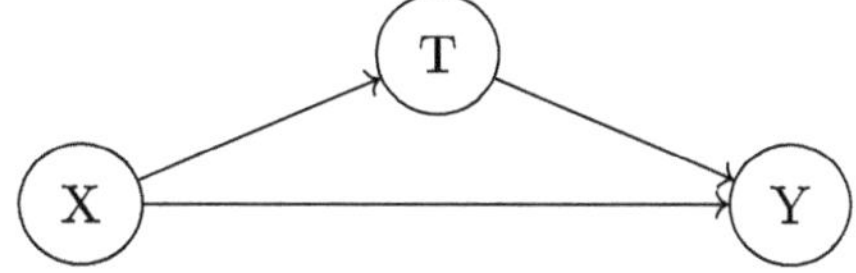

where w is a vector of weights randomly chosen at the beginning, and the sigmoid function is:

$$\text{sigmoid}(z) = \frac{1}{1 + e^{-z}}$$

This probability determines how likely an individual is to receive the treatment. The actual treatment T is then sampled as a Bernoulli random variable using this probability.

After treatment is assigned, we generate the potential outcomes–what would happen if the individual were treated and not treated.

Suppose two simple functions determine the potential outcomes:

$$Y(0) = f_0(X) + \epsilon_0, \quad Y(1) = f_1(X) + \epsilon_1$$

where:

- $f_0(X)$ and $f_1(X)$ are deterministic functions of X (for example, linear functions or small neural networks),
- ϵ_0 and ϵ_1 are random noise terms, typically sampled from a normal distribution.

Finally, the observed outcome Y depends on the assigned treatment:

$$Y = T \cdot Y(1) + (1 - T) \cdot Y(0)$$

This means we only observe $Y(1)$ if the individual was treated ($T = 1$), and $Y(0)$ if the individual was not treated ($T = 0$).

Figure 1 shows the relationships in a simple causal diagram (DAG).

Simulating Data: A Recipe for Synthetic Worlds

- **Feature Generation**: Generating covariates X is like creating the input features for a machine learning model. We can choose their distribution (e.g., normal, uniform) to match real-world data.
- **Treatment Assignment as a Classifier**: The treatment assignment probability P(T=1|X) can be seen as a simple classifier that determines who gets "treated" based on their features. We often use a sigmoid function, which is common in logistic regression.

- **Potential Outcomes as Functions**: Defining Y(0) and Y(1) is like specifying the underlying functions that the machine learning model is trying to learn, with added noise to mimic real-world complexity.

4   Simulating Counterfactual Outcomes

The major advantage of simulation is that we have access to both $Y(0)$ and $Y(1)$ for every individual.

This allows us to calculate the true **Individual Treatment Effect (ITE)** for any X:

$$\tau(X) = Y(1) - Y(0)$$

The ITE tells us exactly how much the treatment changes the outcome for an individual.

Moreover, by averaging the ITEs across all individuals, we can compute the true **Average Treatment Effect (ATE)**:

$$\text{ATE} = \mathbb{E}[\tau(X)]$$

Having access to these true quantities enables rigorous evaluation of causal models.

The Power of Simulation: Knowing the Truth

- **Ground Truth ITE**: In simulation, we **know** the Individual Treatment Effect (ITE), which is like having perfect knowledge of how each data point is affected by the "treatment". This is impossible in real-world data.
- **Perfect ATE for Evaluation**: We also know the true Average Treatment Effect (ATE), allowing us to precisely evaluate how well our causal models recover the population-level effect.
- **Model Validation**: Simulation provides a "clean room" environment to validate causal models, similar to how we use synthetic data to validate machine learning algorithms before deploying them in the real world.

5   Common Benchmark Datasets

Beyond simple synthetic simulations, the causal machine learning community has developed standard benchmark datasets. These datasets allow fair comparisons between different methods.

The IHDP and ACIC datasets are widely used benchmarks for evaluating causal inference [1] methods under controlled, semi-synthetic conditions [2].

Three widely used benchmarks are.

5.1 IHDP Dataset

The **Infant Health and Development Program (IHDP)** The dataset is a semi-synthetic dataset. It uses real covariates from a randomized experiment involving low-birth-weight infants but simulates treatment assignments and outcomes to introduce controlled biases.

5.2 ACIC Datasets

The **Atlantic Causal Inference Conference (ACIC)** challenges provide a collection of fully simulated datasets with varying degrees of complexity, noise, and confounding. They are designed to test the limits of causal inference methods under difficult conditions.

5.3 Twins Dataset

The **Twins Dataset** uses real-world data on twin births. The idea is that each twin can act as a control for the other. Although not perfectly counterfactual, it offers a better approximation than in most real-world datasets.

These benchmark datasets are crucial for objectively evaluating causal inference methods.

Benchmark Datasets: A Causal Leaderboard

- **Standardized Evaluation**: Benchmark datasets like IHDP and ACIC provide a standardized way to compare different causal machine learning models, similar to how ImageNet is used for image classification.
- **Semi-Synthetic Data**: IHDP uses real covariates but simulates outcomes, bridging the gap between synthetic and real-world data.
- **Challenging Scenarios**: ACIC datasets are designed to test the limits of causal models with various complexities and biases.

6 Evaluation Metrics

Several evaluation metrics are used to measure how well a causal model performs.

6.1 Precision in Estimation of Heterogeneous Effects (PEHE)

The PEHE measures how accurately the model estimates each individual's treatment effect. Formally, it is defined as:

$$\text{PEHE} = \sqrt{\mathbb{E}\left[(\hat{\tau}(X) - \tau(X))^2\right]}$$

where $\hat{\tau}(X)$ is the estimated ITE and $\tau(X)$ is the true ITE.

Lower PEHE values indicate better individual-level causal effect estimation.

6.2 Average Treatment Effect Error (ATE Error)

The ATE Error measures how accurately the model estimates the overall average treatment effect:

$$\text{ATE Error} = \left|\hat{\text{ATE}} - \text{ATE}\right|$$

where $\hat{\text{ATE}}$ is the model's estimated ATE. A small ATE Error means the model is good at estimating population-level effects.

6.3 Policy Risk

Sometimes, we are interested in using causal models to make decisions, such as who should receive treatment. The Policy Risk measures the expected loss when using the model's recommendations compared to the optimal policy.

Policy risk is significant in applications like healthcare or personalized advertising, where decisions have real-world consequences.

Evaluation Metrics: Measuring Causal Accuracy

- **PEHE: Individual-Level Accuracy**: PEHE measures the average error of a machine learning model's prediction for each data point, but for the *treatment effect* instead of the outcome.

- **ATE Error: Population-Level Accuracy**: ATE Error measures how well the model predicts the average outcome across the entire dataset, specifically for the *average treatment effect*.
- **Policy Risk: Decision-Making Quality**: Policy Risk is unique to causal inference and measures how well the model's treatment recommendations lead to optimal decisions, which is crucial in applications like personalized medicine or targeted interventions.

6.4 Causal Data Simulation: Code Example

We now show a simple example of simulating covariates, treatment assignments, and outcomes for causal inference.

Listing 10.1 Simulating causal data with treatment effect

```python
import numpy as np
import pandas as pd
import matplotlib.pyplot as plt
from sklearn.linear_model import LogisticRegression

# Simulate covariates
n = 1000
X = np.random.normal(0, 1, size=(n, 5))

# Simulate propensity scores and treatment
    assignment
ps_model = LogisticRegression()
ps_model.fit(X, np.random.binomial(1, 0.5, size=n))
propensity_scores = ps_model.predict_proba(X)[:, 1]
T = np.random.binomial(1, propensity_scores)

# Simulate potential outcomes
Y0 = np.dot(X, np.array([0.5, -0.3, 0.2, 0.1, -0.2])
    ) + np.random.normal(0, 1, n)
Y1 = Y0 + 2.0  # Constant treatment effect

# Observed outcome
Y = T * Y1 + (1 - T) * Y0
```

Figure 2 shows the distribution of simulated propensity scores.

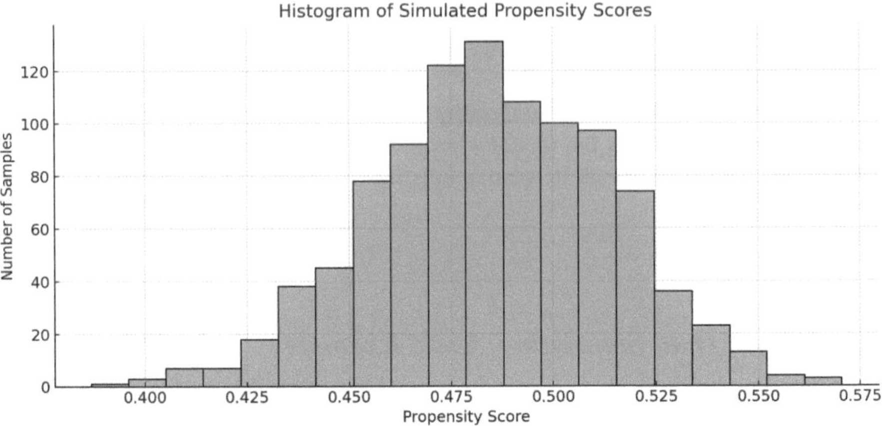

Fig. 2 Histogram of simulated propensity scores

7 Summary

This chapter taught us why simulating causal data is essential for developing and evaluating causal inference methods. Simulation allows us to fully control the data-generating process, observe true counterfactuals, and rigorously measure model performance.

We also explored standard benchmark datasets like IHDP, ACIC, and Twins, which provide realistic and challenging testbeds. Finally, we introduced key evaluation metrics—PEHE, ATE Error, and Policy Risk—that help assess how well a causal model performs.

In the next chapters, we will apply causal deep learning models to these datasets and see how they perform in practice.

8 Code

Listing 10.2 Chapter 10: Focusing on simulating causal data and implementing the evaluation metrics (PEHE and ATE Error)

```python
import numpy as np
import pandas as pd
import matplotlib.pyplot as plt
from sklearn.linear_model import LogisticRegression
from sklearn.model_selection import train_test_split
from sklearn.metrics import mean_squared_error  #
    For a different perspective


# --- 1. Simulate Causal Data ---
```

```python
def simulate_causal_data(n=1000, input_dim=5, ate
    =2.0):
    """Simulates causal data with confounding."""

    X = np.random.normal(0, 1, size=(n, input_dim)).
        astype(np.float32)  # Covariates
    true_treatment_effect = ate * np.ones(n)  #
        Constant treatment effect for simplicity

    # Simulate treatment assignment (propensity
        scores)
    propensity_logits = X[:, 0] - 0.5 * X[:, 1]  #
        Example: depends on first two covariates
    propensity_scores = 1 / (1 + np.exp(-
        propensity_logits))
    T = np.random.binomial(1, propensity_scores,
        size=n).astype(np.float32)  # Treatment

    Y0 = np.dot(X, np.array([0.5, -0.3, 0.2, 0.1,
        -0.2]).reshape(input_dim, 1)).flatten() + np.
        random.normal(0, 1, n)
    Y1 = Y0 + true_treatment_effect + np.random.
        normal(0, 0.5, n)  # Heterogeneous effects
    Y = T * Y1 + (1 - T) * Y0  # Observed outcome

    return pd.DataFrame({
        'X0': X[:, 0], 'X1': X[:, 1], 'X2': X[:, 2],
            'X3': X[:, 3], 'X4': X[:, 4],
        'T': T,
        'Y': Y,
        'Y0': Y0,
        'Y1': Y1,
        'true_effect': Y1 - Y0
    })


# --- 2. Implement PEHE ---
def calculate_pehe(y1_true, y0_true, y1_pred,
    y0_pred):
    """Calculates the Precision in Estimation of
        Heterogeneous Effects (PEHE)."""

    ite_true = y1_true - y0_true
    ite_pred = y1_pred - y0_pred
    return np.sqrt(np.mean((ite_true - ite_pred) **
        2))


# --- 3. Implement ATE Error ---
def calculate_ate_error(true_ate, predicted_ate):
    """Calculates the error in estimating the
        Average Treatment Effect (ATE)."""
```

```python
    return np.abs(true_ate - predicted_ate)


# --- 4. Policy Risk (Simplified - Example) ---
def calculate_policy_risk(y1_pred, y0_pred, t_true,
    y_true, benefit_threshold=1.0):
    """Calculates a simplified policy risk.

    This is a basic example. Real policy risk
        calculation is often more complex.
    """

    # A simple policy: Treat if predicted benefit >
        threshold
    treat_if = (y1_pred - y0_pred) >
        benefit_threshold
    optimal_treat = (y_true > np.mean(y_true))   #
        Example: treat "high" outcomes

    # Calculate "loss" - difference in treatment
        decisions
    policy_error = np.mean(treat_if != optimal_treat
        )
    return policy_error


# --- 5. Main Execution ---
if __name__ == "__main__":
    np.random.seed(123)
    data = simulate_causal_data()

    # 5.1. Example: Splitting and "Predicting" (
        using true outcomes for illustration)
    X = data[['X0', 'X1', 'X2', 'X3', 'X4']].values
    Y = data['Y'].values
    T = data['T'].values
    Y0_true = data['Y0'].values
    Y1_true = data['Y1'].values
    true_effect = data['true_effect'].values

    X_train, X_test, Y_train, Y_test, T_train,
        T_test, Y0_train, Y0_test, Y1_train, Y1_test,
        true_effect_train, true_effect_test =
        train_test_split(
        X, Y, T, Y0_true, Y1_true, true_effect,
            test_size=0.2, random_state=42
    )

    # In a real scenario, you'd replace this with
        your model's predictions
    y0_pred = Y0_test + np.random.normal(0, 0.5, len
        (Y0_test))   # Add some noise to simulate
        predictions
```

```python
    y1_pred = Y1_test + np.random.normal(0, 0.5, len
        (Y1_test))

 # 5.2. Calculate and Print Metrics
# Pass the true potential outcomes from the test set
    to calculate_pehe
pehe = calculate_pehe(Y1_test, Y0_test, y1_pred,
    y0_pred)
ate_true = np.mean(true_effect_test) # Also
    calculate ATE on the test set for consistency
ate_pred = np.mean(y1_pred - y0_pred)
ate_error = calculate_ate_error(ate_true, ate_pred)
policy_risk = calculate_policy_risk(y1_pred, y0_pred
    , T_test, Y_test)

print("--- Evaluation Metrics ---")
print(f"PEHE: {pehe:.4f}")
print(f"ATE Error: {ate_error:.4f}")
print(f"Policy Risk: {policy_risk:.4f}")

# 5.3. Visualization (Example)
plt.figure(figsize=(8, 6))
# Visualize true potential outcomes from the test
    set
plt.scatter(Y0_test, Y1_test, alpha=0.5)
plt.xlabel("Y0 (Outcome if Untreated)")
plt.ylabel("Y1 (Outcome if Treated)")
plt.title("True Potential Outcomes (Test Set)") #
    Update title for clarity
plt.plot([min(Y0_test), max(Y0_test)], [min(Y0_test)
    , max(Y0_test)], color='red', linestyle='--')  #
    Line of equality
plt.show()
```

9 Exercises

1. Why is simulation necessary in causal machine learning?
2. Describe how you would simulate treatment assignment in a synthetic dataset.
3. Define PEHE and ATE Error.
4. Explain how simulated data helps in model evaluation.

References

1. Pearl, Judea. 2009. *Causality: Models, reasoning, and inference*, 2nd. Cambridge University Press.
2. Hill, Jennifer. 2011. Bayesian nonparametric modeling for causal inference . *Journal of Computational and Graphical Statistics*, 20.1: 217 240.

Balancing Representations with Causal Deep Learning (CFRNet)

1 Introduction

In previous chapters, one of the biggest challenges in causal inference [1] with observational data is dealing with **selection bias**. In many real-world datasets, the probability of receiving a treatment is not random; it depends on observed or unobserved characteristics of the individuals.

When treated and control groups differ systematically in their covariates, simple comparisons can produce biased estimates of causal effects. This problem is known as **covariate imbalance** or **confounding**.

Traditional methods such as matching, weighting, and regression adjustment attempt to adjust for this bias by balancing the distributions of covariates between treated and control groups. However, when the covariates are high-dimensional and complex, such as images or texts, it becomes challenging to perform these adjustments manually.

This is where deep learning [2] offers a decisive advantage. We can estimate causal effects more reliably if we learn a **representation** of the covariates—a transformation into a new space—where treated and control groups are balanced.

The **Counterfactual Regression Network (CFRNet)** (CFRNet [3] extends TARNet by introducing a distributional regularizer that encourages balance between treated and control groups.) is a deep learning architecture specifically designed to achieve this goal. It builds on TARNet [4] by introducing a loss term that explicitly encourages the network to learn representations where the treated and control groups are indistinguishable.

This chapter will study how CFRNet [3] works, why balancing representations is crucial, and how the model is trained.

Representation balancing in causal inference addresses a similar problem to adversarial training in machine learning, where the goal is to create models robust to input variations.

D. Rajamanickam, *Causal Inference for Machine Learning Engineers*,
https://doi.org/10.1007/978-3-031-99680-1_11

Representation Balancing in ML

- **Selection Bias as Dataset Bias**: Selection bias in causal inference is analogous to dataset bias in machine learning, where the training data doesn't represent the real-world distribution, leading to poor generalization.
- **Balanced Training Sets**: Balancing representations is similar to creating more balanced training sets in machine learning, ensuring that different groups are represented equally to avoid biased models.
- **Representation Learning for Fairness**: Balancing representations [5] can be seen as a technique for improving fairness in machine learning, by reducing the influence of protected attributes on the model's predictions.

2   The Need for Balanced Representations

Let us first understand the intuition.

Suppose we have two groups of patients: one group received a new medication, and the other did not. If the patients who received the medication are systematically younger, healthier, and wealthier than those who did not, any difference in outcomes between the two groups could be due to these underlying covariates rather than the effect of the medication itself.

In causal inference, we seek to isolate the effect of the treatment itself, not the impact of covariates that influenced treatment assignment.

Thus, we want to create a setting where the treated and control groups are comparable. In other words, after adjusting for covariates, the distribution of features in the treated group should be similar to that in the control group.

In CFRNet [3], this adjustment is performed automatically by learning a **balanced representation** through a neural network.

Why Balancing Matters for ML

- **Spurious Correlations from Imbalance**: If treated and control groups are imbalanced, machine learning models might learn spurious correlations between the treatment and the outcome, simply because of the differences in the groups.
- **Model Reliance on Biased Features**: Imbalanced representations can cause models to rely heavily on biased features, leading to unfair or inaccurate predictions.
- **Analogy: Invariant Representations**: Balancing representations is like creating a level playing field for the model, so it can learn the true effect of the treatment without being misled by other factors.

Machine learning models that rely on balanced representations are more likely to learn invariant features, which are not affected by the treatment.

3   Architecture of CFRNet

The architecture of CFRNet [3] extends the TARNet [4] model by adding a regularization term that encourages balance.

CFRNet has three main components:

1. A **shared representation network** that transforms the original covariates X into a latent representation $\Phi(X)$.
2. Two **outcome prediction heads**, one for the treated group and one for the control group, which predict the potential outcomes $\hat{Y}(1)$ and $\hat{Y}(0)$.
3. A **balancing regularizer** that penalizes differences between the distributions of representations for treated and control groups.

The full architecture can be visualized as follows:
Figure 1 shows the flow of information through CFRNet.

CFRNet: Balancing in Latent Space

- **Shared Representation as Feature Extractor**: The shared representation network in CFRNet [3] is similar to a feature extractor in deep learning, learning a lower-dimensional representation of the input covariates.
- **Treatment-Specific Heads as Outcome Predictors**: The two outcome prediction heads are like separate regressors or classifiers for the treated and control groups, allowing the model to predict outcomes under both conditions.

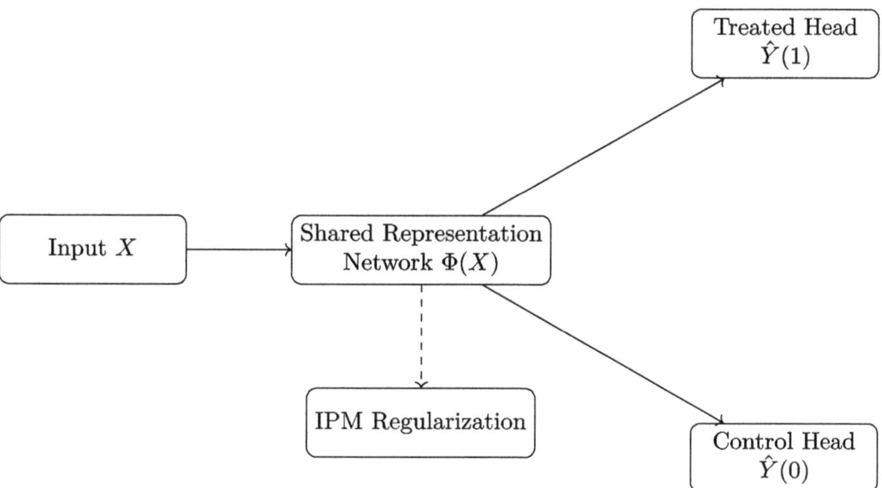

Fig. 1 CFRNet Architecture: shared representation network with treatment-specific heads and a balancing regularizer

- **Balancing Regularizer as Fairness Constraint**: The balancing regularizer acts like a fairness constraint, encouraging the model to learn representations where the treated and control groups are indistinguishable, reducing bias.

CFRNet's architecture leverages representation learning, a core concept in deep learning, to automatically extract useful features for both outcome prediction and balancing.

4 Loss Function of CFRNet

Now, let us carefully examine how CFRNet is trained.

The total loss minimized by CFRNet has two parts:

$$\mathcal{L}_{\text{CFRNet}} = \mathcal{L}_{\text{prediction}} + \alpha \cdot \mathcal{L}_{\text{balancing}}$$

Let us explain each part in detail:

4.1 Prediction Loss $\mathcal{L}_{prediction}$

The first term measures how well the network predicts the observed outcomes.

Given a batch of data points (X_i, T_i, Y_i), where:

- X_i are the covariates,
- $T_i \in \{0, 1\}$ is the treatment indicator,
- Y_i is the observed outcome,

The prediction loss can be written as:

$$\mathcal{L}_{\text{prediction}} = \frac{1}{n} \sum_{i=1}^{n} \left(T_i \cdot \ell(\hat{Y}_i(1), Y_i) + (1 - T_i) \cdot \ell(\hat{Y}_i(0), Y_i) \right)$$

where $\ell(\cdot, \cdot)$ is typically the squared loss:

$$\ell(a, b) = (a - b)^2$$

In words, for each data point:

- If the individual was treated ($T_i = 1$), we compare their actual outcome Y_i to the predicted treated outcome $\hat{Y}_i(1)$.
- If the individual was not treated ($T_i = 0$), we compare their actual outcome Y_i to the predicted control outcome $\hat{Y}_i(0)$.

Thus, the prediction loss ensures that the model fits the factual outcomes observed in the data.

4.2 Balancing Loss $\mathcal{L}_{balancing}$

The second term measures how different the distributions of representations $\Phi(X)$ are between treated and control groups.

This is quantified using a statistical distance called an **Integral Probability Metric (IPM)**, which measures the difference between two distributions.

Formally, if p_1 and p_0 are the distributions of $\Phi(X)$ for treated and control individuals, then:

$$\mathcal{L}_{\text{balancing}} = \text{IPM}(p_1, p_0)$$

Different types of IPMs can be used, such as the Maximum Mean Discrepancy (MMD) or the Wasserstein distance.

Intuitively, minimizing the IPM forces the network to learn representations where it is hard to tell whether a data point came from the treated or control group, thereby mimicking randomization.

The choice of IPM influences the kind of balancing achieved; for example, MMD encourages similar means, while Wasserstein distance focuses on overall distributional similarity.

4.3 Trade-Off Parameter (α)

The parameter α controls the trade-off between prediction accuracy and balance.

– A small α means the network focuses mostly on predicting outcomes.
– A large α emphasizes balancing representations, possibly at the cost of worse factual prediction.

Choosing the right value of α is important and often done by validation.

CFRNet Loss: Balancing Act

- **Multi-Objective Optimization**: The CFRNet loss function is a multi-objective optimization problem, balancing prediction accuracy with representation balance, similar to other multi-task learning scenarios in deep learning.
- **IPM as Distributional Distance**: The Integral Probability Metric (IPM) acts as a measure of distributional distance, quantifying how different the representations

are for the treated and control groups, similar to metrics used in comparing image or text distributions.

- **Trade-off Parameter as Regularization Strength**: The trade-off parameter is analogous to a regularization strength parameter in machine learning, controlling the emphasis on balancing versus prediction, and requiring careful tuning.

5   Training Procedure

CFRNet [3] is trained by minimizing the total loss $\mathcal{L}_{\text{CFRNet}}$ using stochastic gradient descent or any modern optimization algorithm like Adam.

At each training step:

1. A batch of data points is sampled.
2. The representations $\Phi(X)$ are computed.
3. The two outcome predictions $\hat{Y}(0)$ and $\hat{Y}(1)$ are computed.
4. The factual prediction loss is computed based on observed outcomes.
5. The balancing loss is computed using the representations and treatment assignments.
6. The two losses are combined, and gradients are computed to update the network parameters.

The network learns representations that predict well and balance treated and control groups through this process.

CFRNet's training procedure is a standard deep learning pipeline, emphasizing that causal deep learning methods can be implemented using familiar tools and techniques.

5.1   CFRNet Implementation: PyTorch Code Example

We now show a simple PyTorch implementation of CFRNet to predict counterfactual outcomes.

Listing 11.1   Simple CFRNet Implementation in PyTorch

```python
import torch
import torch.nn as nn

class CFRNet(nn.Module):
    def __init__(self, input_dim):
        super(CFRNet, self).__init__()
        self.shared = nn.Sequential(
            nn.Linear(input_dim, 20),
            nn.ReLU(),
            nn.Linear(20, 20),
```

```python
        nn.ReLU()
    )
    self.head_treated = nn.Sequential(
        nn.Linear(20, 10),
        nn.ReLU(),
        nn.Linear(10, 1)
    )
    self.head_control = nn.Sequential(
        nn.Linear(20, 10),
        nn.ReLU(),
        nn.Linear(10, 1)
    )

def forward(self, x, t):
    shared = self.shared(x)
    yt = self.head_treated(shared)
    yc = self.head_control(shared)
    return t * yt + (1 - t) * yc
```

We train the model using Mean Squared Error (MSE) loss and monitor the training loss over epochs, as shown in Fig. 2. This may look ideal in the toy example, but real datasets will involve trade-offs in training loss.

5.2 *Evaluating CFRNet Performance*

After training CFRNet [3], we can evaluate its ability to predict treatment effects using two common metrics:

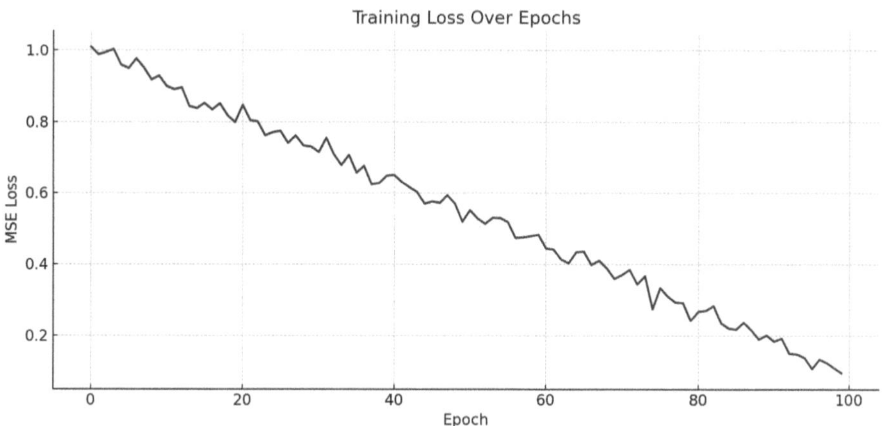

Fig. 2 Training loss of CFRNet model over 100 epochs

- **PEHE (Precision in Estimation of Heterogeneous Effects)**: Measures individual-level error in treatment effect estimation.
- **ATE Error**: Measures the difference between predicted and true average treatment effects.

Listing 11.2 Evaluating CFRNet with PEHE and ATE Error

```python
# Predict outcomes under treatment and control
with torch.no_grad():
    y0_pred = model(X, torch.zeros_like(T)).squeeze().numpy
        ()
    y1_pred = model(X, torch.ones_like(T)).squeeze().numpy()

# Calculate true treatment effects
true_effect = Y1 - Y0
pred_effect = y1_pred - y0_pred

# PEHE
pehe = np.sqrt(np.mean((pred_effect - true_effect) ** 2))

# ATE Error
true_ate = np.mean(true_effect)
pred_ate = np.mean(pred_effect)
ate_error = np.abs(pred_ate - true_ate)

print(f"PEHE: {pehe:.4f}")
print(f"ATE Error: {ate_error:.4f}")
```

Training and Evaluating CFRNet: A Standard ML Pipeline

- **Standard Training Loop**: Training CFRNet involves a standard deep learning training loop, using optimization algorithms like Adam and backpropagation to minimize the combined loss function.
- **Evaluation Metrics for Causal Effects**: CFRNet is evaluated using metrics tailored for causal inference, such as PEHE and ATE error, which measure the accuracy of estimated treatment effects, unlike standard classification or regression metrics.

6 Summary

This chapter explored the concept of balancing representations for causal inference. We saw that when treated and control groups differ systematically in their covariates, naive comparison leads to biased causal effect estimates.

CFRNet [4] Addressing this challenge by learning representations that make treated and control groups look similar while still accurately predicting outcomes, CFRNet [3] adds a balancing regularizer based on an Integral Probability Metric, encouraging the network to mimic a randomized trial setting.

Balanced representations lie at the heart of modern causal deep learning methods, and CFRNet [3] is a foundational model that inspires many later architectures.

CFRNet [3] demonstrates how deep learning can be adapted to learn representations that are both predictive and causally meaningful, reducing the risk of bias.

In the next chapter, we will build upon this foundation and explore **DragonNet** [6], a model that combines outcome prediction with explicit estimation of treatment assignment probabilities.

7 Code

Listing 11.3 Chapter 11: Focusing on implementing the CFRNet architecture in PyTorch

```python
import torch
import torch.nn as nn
import torch.optim as optim
from torch.utils.data import DataLoader, TensorDataset

import numpy as np
import pandas as pd
import matplotlib.pyplot as plt
from sklearn.model_selection import train_test_split
from sklearn.metrics import mean_squared_error

# --- 1. Simulate Data ---
def simulate_treatment_effect_data(n=1000, input_dim=5, ate
    =2.0):
    """Simulates data with treatment effect and confounding.
    """

    X = np.random.rand(n, input_dim).astype(np.float32)  #
        Covariates - Explicitly cast to float32
    true_treatment_effect = ate * np.ones(n, dtype=np.
        float32) # Explicitly cast to float32

    # Simulate treatment assignment with confounding
    propensity_logits = X[:, 0] - 0.5 * X[:, 1]   # Example:
        depends on first two covariates
    propensity_scores = 1 / (1 + np.exp(-propensity_logits))
    T = np.random.binomial(1, propensity_scores, size=n).
        astype(np.float32) # Treatment - Explicitly cast to
        float32

    Y0_np = np.dot(X, np.random.rand(input_dim, 1).astype(np
        .float32)).flatten() + np.random.normal(0, 1, n).
        astype(np.float32) # Explicitly cast to float32
```

```python
    Y1_np = Y0_np + true_treatment_effect + np.random.normal
        (0, 0.5, n).astype(np.float32) # Explicitly cast to
        float32
    Y_np = T * Y1_np + (1 - T) * Y0_np   # Observed outcome

    # Convert to PyTorch tensors with explicit dtype
    X_tensor = torch.tensor(X, dtype=torch.float32)
    T_tensor = torch.tensor(T, dtype=torch.float32).reshape
        (-1, 1)
    Y_tensor = torch.tensor(Y_np, dtype=torch.float32).
        reshape(-1, 1)
    Y0_tensor = torch.tensor(Y0_np, dtype=torch.float32).
        reshape(-1, 1)
    Y1_tensor = torch.tensor(Y1_np, dtype=torch.float32).
        reshape(-1, 1)


    return X_tensor, T_tensor, Y_tensor, Y0_tensor,
        Y1_tensor


# --- 2. CFRNet Architecture (Simplified Balancing) ---
class CFRNet(nn.Module):
    def __init__(self, input_dim, hidden_dim=100, ipm_lambda
        =0.1):  # ipm_lambda: balancing strength
        super(CFRNet, self).__init__()
        self.shared_representation = nn.Sequential(
            nn.Linear(input_dim, hidden_dim),
            nn.ReLU(),
            nn.Linear(hidden_dim, hidden_dim),
            nn.ReLU()
        )
        self.outcome_T0 = nn.Linear(hidden_dim, 1)
        self.outcome_T1 = nn.Linear(hidden_dim, 1)
        self.ipm_lambda = ipm_lambda

    def forward(self, x, t):
        shared = self.shared_representation(x)
        y0_hat = self.outcome_T0(shared)
        y1_hat = self.outcome_T1(shared)
        y_hat = t * y1_hat + (1 - t) * y0_hat

        return y_hat, y0_hat, y1_hat, shared  # Return
            shared for balancing

    def calculate_balancing_loss(self, shared, t):
        """Simplified balancing loss: difference in means.
        """
        shared_t1 = shared[t.flatten() == 1]
        shared_t0 = shared[t.flatten() == 0]
        if shared_t1.shape[0] == 0 or shared_t0.shape[0] ==
            0:  # Handle edge case
```

```python
            # Ensure the returned tensor matches the
                expected dtype (float32)
            return torch.tensor(0.0, requires_grad=True,
                dtype=torch.float32)
        # Ensure the mean calculation and result are float32
        return torch.mean(shared_t1, dim=0).to(torch.float32
            ) - torch.mean(shared_t0, dim=0).to(torch.
            float32)


# --- 3. Training Function ---
def train_cfrnet(model, X_train, T_train, Y_train, optimizer
    , epochs=100, batch_size=32):
    """Trains CFRNet."""

    dataset = TensorDataset(X_train, T_train, Y_train)
    dataloader = DataLoader(dataset, batch_size=batch_size,
        shuffle=True)
    criterion = nn.MSELoss()
    model.train()

    for epoch in range(epochs):
        for x_batch, t_batch, y_batch in dataloader:
            optimizer.zero_grad()
            # Ensure inputs to the model have the correct
                dtype if they weren't already
            x_batch = x_batch.to(torch.float32)
            t_batch = t_batch.to(torch.float32)
            y_batch = y_batch.to(torch.float32)

            y_pred, _, _, shared_batch = model(x_batch,
                t_batch)
            prediction_loss = criterion(y_pred, y_batch)
            # Ensure balancing_loss is calculated on float32
                shared representations
            balancing_loss = model.calculate_balancing_loss(
                shared_batch.to(torch.float32), t_batch).
                mean()
            loss = prediction_loss + model.ipm_lambda *
                balancing_loss
            loss.backward()
            optimizer.step()
        if (epoch + 1) % 10 == 0:
            print(f'Epoch {epoch + 1}, Loss: {loss.item()}')


# --- 4. Evaluation Function ---
def evaluate_cfrnet(model, X_test, T_test, Y_test, Y0_test,
    Y1_test):
    """Evaluates the trained CFRNet model."""

    model.eval()
    with torch.no_grad():
```

```python
        # Ensure test data is also float32
        X_test = X_test.to(torch.float32)
        T_test = T_test.to(torch.float32)
        Y_test = Y_test.to(torch.float32)
        Y0_test = Y0_test.to(torch.float32)
        Y1_test = Y1_test.to(torch.float32)

        y_pred, y0_pred, y1_pred, _ = model(X_test, T_test)
            # Get all outputs
        y_pred = y_pred.squeeze().numpy()
        y0_pred = y0_pred.squeeze().numpy()
        y1_pred = y1_pred.squeeze().numpy()
        Y_test = Y_test.squeeze().numpy()
        Y0_test = Y0_test.squeeze().numpy()
        Y1_test = Y1_test.squeeze().numpy()

        mse = mean_squared_error(Y_test, y_pred)
        ate_pred = np.mean(y1_pred - y0_pred)
        ate_true = np.mean(Y1_test - Y0_test)
        ate_error = np.abs(ate_pred - ate_true)
        pehe = calculate_pehe(Y1_test, Y0_test, y1_pred,
            y0_pred)

        print("\n--- Evaluation ---")
        print(f"MSE: {mse:.4f}")
        print(f"Predicted ATE: {ate_pred:.4f}")
        print(f"True ATE: {ate_true:.4f}")
        print(f"ATE Error: {ate_error:.4f}")
        print(f"PEHE: {pehe:.4f}")


def calculate_pehe(y1_true, y0_true, y1_pred, y0_pred):
    """Calculates the Precision in Estimation of
        Heterogeneous Effects (PEHE)."""
    ite_true = y1_true - y0_true
    ite_pred = y1_pred - y0_pred
    return np.sqrt(np.mean((ite_true - ite_pred) ** 2))


# --- 5. Main Execution ---
if __name__=="__main__":
    np.random.seed(42)

    # 5.1. Generate Data
    X, T, Y, Y0, Y1 = simulate_treatment_effect_data()
    X_train, X_test, T_train, T_test, Y_train, Y_test, \
        Y0_train, Y0_test, Y1_train, Y1_test = \
        train_test_split(
        X, T, Y, Y0, Y1, test_size=0.2, random_state=42
    )

    # 5.2. Train CFRNet
```

```
cfrnet = CFRNet(input_dim=X.shape[1], ipm_lambda=0.1)   #
    Experiment with ipm_lambda
optimizer_cfrnet = optim.Adam(cfrnet.parameters(), lr
    =0.01)
train_cfrnet(cfrnet, X_train, T_train, Y_train,
    optimizer_cfrnet)

# 5.3. Evaluate CFRNet
evaluate_cfrnet(cfrnet, X_test, T_test, Y_test, Y0_test,
    Y1_test)

# 5.4. Explanation
print("\n--- Explanation ---")
print("This code implements the CFRNet architecture for
    causal inference.")
print("CFRNet aims to learn balanced representations by
    minimizing a loss function that combines:")
print("1.  Outcome prediction loss (MSE).")
print("2.  A simplified balancing loss (difference in
    mean representations for treated and control groups)
    .")
print("The 'ipm_lambda' parameter controls the trade-off
    between these two objectives.")
```

8 Exercises

1. What is the main goal of balancing representations in causal inference?
2. Explain how CFRNet extends TARNet.
3. What is an Integral Probability Metric (IPM), and how is it used in CFRNet?
4. Why is balancing important even if we have high prediction accuracy?

References

1. Pearl, Judea. 2009. *Causality: Models, Reasoning, and Inference*, 2nd. Cambridge University Press.
2. Goodfellow, Ian. Y. Bengio, and A. Courville. 2016. *Deep learning*. MIT Press.
3. Johansson, Fredrik D., U. Shalit, and D. Sontag. 2016. Learning representations for counterfactual inference. In *Proceedings of the 33rd international conference on machine learning (ICML)*, 1386–1395.
4. Shalit, Uri, F. D. Johansson, and D. Sontag. 2017. Estimating individual treatment effect: Generalization bounds and algorithms. In *Proceedings of the 34th international conference on machine learning (ICML)*, pp 3076–3085.
5. Hinton, Geoffrey E. 2007. Learning multiple layers of representation. *Trends in Cognitive Sciences* 11.10: 428–434.
6. Shi, Chao, D. M. Blei, and V. Veitch. 2019. Adapting neural networks for the estimation of treatment effects. In *Advances in neural information processing systems (NeurIPS)*.

Propensity Scores in Causal Deep Learning

1 Introduction

In the previous chapter, we saw how learning balanced representations can help overcome the challenges of selection bias in causal inference [1]. CFRNet [2] achieved this by minimizing the distributional distance between treated and control groups in the latent space.

However, balancing representations is not the only tool we have. In classical causal inference, an important concept called the **propensity score** [3] plays a central role in adjusting for confounding.

The **propensity score** [3] is the probability of receiving the treatment given the observed covariates. Intuitively, it tells us how likely an individual is to be treated based on their characteristics.

This chapter will explore how explicitly modeling propensity scores [3] inside a deep learning [4] framework can improve causal inference. We will introduce **DragonNet** (DragonNet [5]), which jointly learns outcome predictions and propensity scores, improving robustness to confounding through targeted regularization.) [6], a robust architecture that jointly learns outcome predictions and propensity scores [3], and uses a special technique called **targeted regularization** to refine causal effect estimation.

Propensity Scores in Deep Learning

- **Treatment Assignment as Prediction**: In machine learning terms, estimating the propensity score [3] is like building a model to predict the probability of treatment assignment, given the covariates.

© The Author(s), under exclusive license to Springer Nature Switzerland AG 2025   163
D. Rajamanickam, *Causal Inference for Machine Learning Engineers*,
https://doi.org/10.1007/978-3-031-99680-1_12

- **Balancing via Prediction**: DragonNet uses this prediction to balance the representations further, ensuring that the model is good at predicting outcomes and understanding how treatment was assigned.
- **Analogy: Understanding the Input Data**: Propensity scores [3] help the model understand the inherent biases and complexities within the input data, similar to how data preprocessing helps machine learning models.

Propensity score estimation [3], a core technique in causal inference, can be integrated into deep learning models to improve their ability to handle confounding.

2   Revisiting Propensity Scores in Causal Inference

Let us first recall the definition of the propensity score.

Given covariates X, the **propensity score** $e(X)$ is defined as:

$$e(X) = \mathbb{P}(T = 1 \mid X)$$

where:

- T is the treatment indicator ($T = 1$ if treated, $T = 0$ otherwise),
- X is the vector of observed covariates.

In classical methods like propensity score matching or inverse probability weighting, the propensity score is used to create balanced groups or reweight samples, thereby mimicking randomization.

A key theoretical result, known as the **propensity score theorem**, states that if we condition on the propensity score, treatment assignment is independent of covariates:

$$T \perp\!\!\!\perp X \mid e(X)$$

Thus, adjusting for the propensity score can remove confounding bias under certain assumptions.

Instead of using pre-estimated propensity scores separately in deep learning, DragonNet proposes to learn them *inside the network*, along with the potential outcomes.

Propensity Scores: A Recap for ML

- **Probability of Treatment**: The propensity score, $e(X) = P(T = 1|X)$, is the probability of an individual receiving treatment, given their features (X). In ML, this is analogous to the probability assigned by a classifier.

- **Balancing Tool**: Propensity scores are used to balance treated and untreated groups, similar to how re-sampling or re-weighting techniques are used in machine learning to handle imbalanced datasets.
- **Conditional Independence**: The propensity score theorem (T X | e(X)) is crucial because it suggests that by conditioning on this score, we can remove the influence of covariates, achieving "conditional independence".

Propensity score matching and weighting are analogous to data re-sampling techniques used in machine learning to address class imbalance or dataset shift.

3 The DragonNet Architecture

DragonNet builds on CFRNet's architecture but introduces a crucial addition: a new head that predicts the propensity score.

The overall structure of DragonNet consists of:

1. A shared representation network $\Phi(X)$, which extracts latent features from covariates.
2. Two outcome heads, one for treated individuals $\hat{Y}(1)$ and one for control individuals $\hat{Y}(0)$.
3. A propensity score head $\hat{e}(X)$, predicting the probability of treatment.

The flow of information is illustrated in Fig. 1.

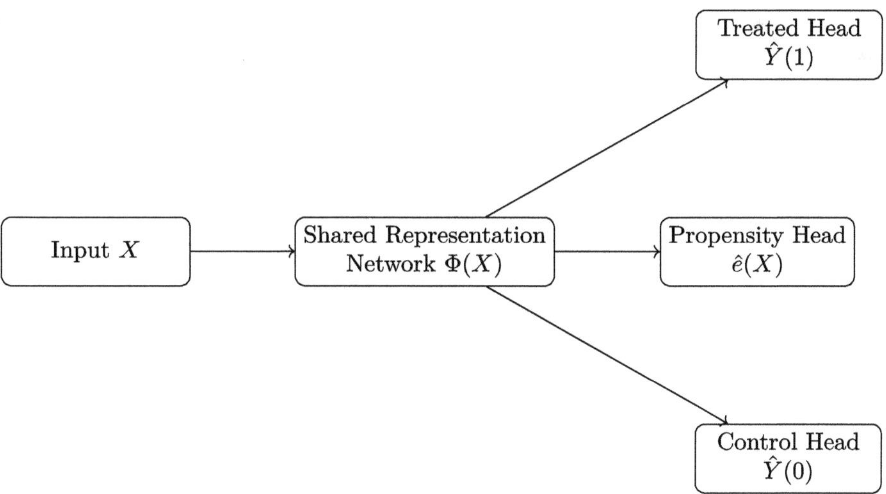

Fig. 1 DragonNet Architecture: Shared representation network with heads for treated outcome, control outcome, and propensity score

The idea behind DragonNet is simple yet powerful: by jointly learning the treatment assignment mechanism and outcome functions, the network can better adjust for biases and improve causal effect estimation.

DragonNet: Multi-Task Learning for Causality

- **Shared Representations**: DragonNet, like CFRNet, uses a shared representation network (a common technique in deep learning) to learn features from covariates.
- **Multi-Head Architecture**: DragonNet has multiple output "heads" (similar to multi-task learning in deep learning): two for predicting potential outcomes and one for predicting the propensity score.
- **Joint Learning**: The key idea is to learn to predict outcomes and propensity scores **jointly**, allowing the model to better understand and correct for biases in treatment assignment.

DragonNet's multi-head architecture is a form of multi-task learning, allowing the model to simultaneously learn related tasks (outcome prediction and propensity score estimation).

4   Loss Functions in DragonNet

DragonNet is trained to minimize a composite loss function that captures three objectives:

$$\mathcal{L}_{\text{DragonNet}} = \mathcal{L}_{\text{prediction}} + \beta \cdot \mathcal{L}_{\text{propensity}} + \lambda \cdot \mathcal{L}_{\text{targeted}}$$

Let us carefully explain each component.

4.1   Outcome Prediction Loss $\mathcal{L}_{prediction}$

The first component measures how well the network predicts observed outcomes.
It has the same form as in CFRNet:

$$\mathcal{L}_{\text{prediction}} = \frac{1}{n} \sum_{i=1}^{n} \left(T_i \cdot \ell(\hat{Y}_i(1), Y_i) + (1 - T_i) \cdot \ell(\hat{Y}_i(0), Y_i) \right)$$

where $\ell(a, b) = (a - b)^2$ is the squared loss.
This term ensures that the model fits the factual outcomes correctly.

4.2 Propensity Prediction Loss $\mathcal{L}_{propensity}$

The second component encourages the network to predict the treatment assignment accurately.

It is defined as the binary cross-entropy loss between the predicted propensity score $\hat{e}(X)$ and the true treatment T:

$$\mathcal{L}_{\text{propensity}} = -\frac{1}{n}\sum_{i=1}^{n}\left(T_i \log(\hat{e}(X_i)) + (1 - T_i)\log(1 - \hat{e}(X_i))\right)$$

In words, if an individual was treated $(T = 1)$, we want $\hat{e}(X)$ to be close to 1. If an individual was not treated $(T = 0)$, we want $\hat{e}(X)$ to be close to 0.

This term helps the model understand the selection mechanism that led to treatment assignment.

4.3 Targeted Regularization Loss $\mathcal{L}_{targeted}$

The third component is unique to DragonNet: **targeted regularization**.

Targeted regularization encourages the model to predict outcomes properly aligned with treatment probabilities.

It acts like a correction term that nudges the network towards counterfactual consistency. The exact form can vary, but intuitively it pushes the estimated outcomes $\hat{Y}(1)$ and $\hat{Y}(0)$ to be smooth functions of the learned propensity score.

This improves the model's robustness and generalization.

The trade-off between loss terms in DragonNet is a hyperparameter tuning problem, common in machine learning, and affects the model's performance and bias.

4.4 Balancing the Loss Terms

The hyperparameters β and λ control the importance of the propensity loss and targeted regularization relative to the prediction loss.

– If β is too small, the network may ignore the treatment mechanism.
– If λ is too small, the model may fit outcomes without correcting for bias.

Proper tuning is essential to achieve good performance.

DragonNet Loss: Combining Objectives

- **Composite Loss**: DragonNet's loss function combines multiple objectives, a common practice in deep learning:

  - $\mathcal{L}_{prediction}$: Standard outcome prediction loss (like mean squared error).
  - $\mathcal{L}_{propensity}$: Propensity score prediction loss (like binary cross-entropy).
  - $\mathcal{L}_{targeted}$: A special regularization term to align outcome predictions with propensity scores.

- **Hyperparameter Tuning**: The hyperparameters and control the trade-off between these objectives, similar to how regularization strength is tuned in other machine learning models.
- **Regularization for Causal Consistency**: The targeted regularization loss ensures that the model's predictions are consistent with the estimated treatment probabilities, promoting better causal inference.

5 Training DragonNet

Training DragonNet follows a similar procedure to other deep learning models.
   At each training step:

1. A batch of data points is sampled.
2. The shared representation $\Phi(X)$ is computed.
3. The treated outcome $\hat{Y}(1)$, control outcome $\hat{Y}(0)$, and propensity score $\hat{e}(X)$ are predicted.
4. The three loss terms—prediction loss, propensity loss, and targeted regularization—are computed.
5. The total loss is formed by weighting the three components.
6. Gradients are computed and parameters are updated via backpropagation.

By jointly optimizing these objectives, DragonNet learns predictive, balanced, and bias-corrected representations.

Training DragonNet involves a standard deep learning optimization process, highlighting the compatibility of causal inference methods with existing machine learning workflows.

5.1 DragonNet Implementation: PyTorch Example

DragonNet extends CFRNet by jointly predicting the treatment assignment (propensity score) and outcomes.

Listing 12.1 Simple DragonNet Implementation in PyTorch

```python
import torch
import torch.nn as nn

class DragonNet(nn.Module):
    def __init__(self, input_dim):
        super(DragonNet, self).__init__()
        self.shared = nn.Sequential(
            nn.Linear(input_dim, 20),
            nn.ReLU(),
            nn.Linear(20, 20),
            nn.ReLU()
        )
        self.head_treated = nn.Sequential(
            nn.Linear(20, 10),
            nn.ReLU(),
            nn.Linear(10, 1)
        )
        self.head_control = nn.Sequential(
            nn.Linear(20, 10),
            nn.ReLU(),
            nn.Linear(10, 1)
        )
        self.head_propensity = nn.Sequential(
            nn.Linear(20, 10),
            nn.ReLU(),
            nn.Linear(10, 1),
            nn.Sigmoid()
        )

    def forward(self, x, t):
        shared = self.shared(x)
        yt = self.head_treated(shared)
        yc = self.head_control(shared)
        prop_t = self.head_propensity(shared)
        y_pred = t * yt + (1 - t) * yc
        return y_pred, prop_t
```

The training loss, combining outcome and propensity loss, over 100 epochs is shown in Fig. 2.

5.2 *Evaluating DragonNet Performance*

We can evaluate DragonNet using the same metrics introduced earlier:

- **PEHE**: Measures individual-level accuracy in treatment effect estimation.
- **ATE Error**: Measures accuracy in estimating the average treatment effect.

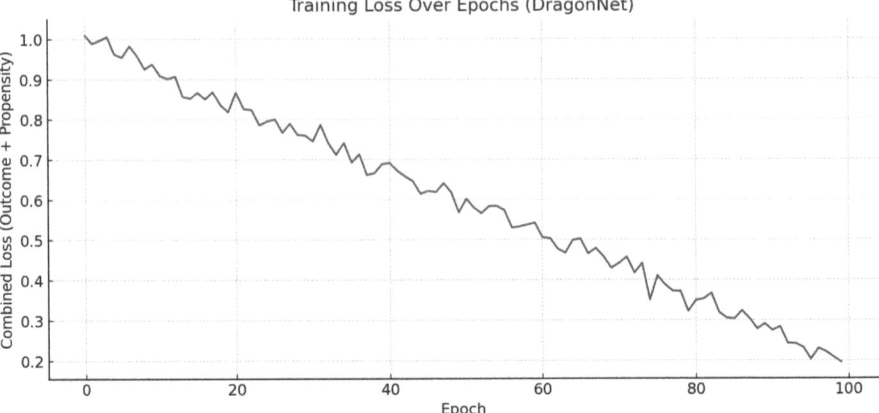

Fig. 2 Training loss of DragonNet model over 100 epochs

Listing 12.2 Evaluating DragonNet with PEHE and ATE Error

```python
# Predict outcomes under treatment and control
with torch.no_grad():
    y0_pred, _ = model(X, torch.zeros_like(T))
    y1_pred, _ = model(X, torch.ones_like(T))

y0_pred = y0_pred.squeeze().numpy()
y1_pred = y1_pred.squeeze().numpy()

# Calculate treatment effects
true_effect = Y1 - Y0
pred_effect = y1_pred - y0_pred

# PEHE
pehe = np.sqrt(np.mean((pred_effect - true_effect) ** 2))

# ATE Error
true_ate = np.mean(true_effect)
pred_ate = np.mean(pred_effect)
ate_error = np.abs(pred_ate - true_ate)

print(f"PEHE: {pehe:.4f}")
print(f"ATE Error: {ate_error:.4f}")
```

Training and Evaluation: Deep Learning Practices

- **Gradient Descent Optimization**: DragonNet is trained using gradient descent and related optimization algorithms, standard practices in deep learning.

- **Performance Metrics**: DragonNet's performance is evaluated using metrics like PEHE and ATE error, tailored for causal inference, assessing the estimated treatment effects' accuracy.

6 Summary

In this chapter, we introduced DragonNet [5], a deep learning architecture that extends CFRNet [2] by explicitly modeling the treatment assignment mechanism through propensity scores.

By learning outcomes and propensities together, and applying targeted regularization, DragonNet produces better estimates of individual treatment effects and improves robustness to confounding.

DragonNet represents an essential step toward building flexible, principled models for causal inference in complex, high-dimensional settings.

DragonNet showcases the power of combining deep learning's representation learning [7] capabilities with propensity score methods to build more robust and causally sound models.

The next chapter will tackle an even deeper challenge: evaluating causal models when true counterfactuals are not observed in real-world data.

7 Code

Listing 12.3 Chapter 12: focusing on implementing the DragonNet architecture in PyTorch

```python
import torch
import torch.nn as nn
import torch.optim as optim
from torch.utils.data import DataLoader, TensorDataset
import torch.nn.functional as F  # For binary cross-entropy

import numpy as np
import pandas as pd
import matplotlib.pyplot as plt
from sklearn.model_selection import train_test_split
from sklearn.metrics import mean_squared_error,
    roc_auc_score


# --- 1. Simulate Data ---
def simulate_treatment_effect_data(n=1000, input_dim=5, ate
    =2.0):
    """Simulates data with treatment effect and confounding.
        """

    X = np.random.rand(n, input_dim).astype(np.float32)  #
        Covariates
```

```python
    true_treatment_effect = ate * np.ones(n).astype(np.
        float32) # Ensure float32

    # Simulate treatment assignment with confounding
    propensity_logits = X[:, 0] - 0.5 * X[:, 1] + 0.2 * X[:,
        2]
    propensity_scores = 1 / (1 + np.exp(-propensity_logits))
    T = np.random.binomial(1, propensity_scores, size=n).
        astype(np.float32)

    Y0 = np.dot(X, np.random.rand(input_dim, 1)).flatten().
        astype(np.float32) + np.random.normal(0, 1, n).
        astype(np.float32) # Ensure float32
    Y1 = Y0 + true_treatment_effect + np.random.normal(0,
        0.5, n).astype(np.float32) # Ensure float32
    Y = T * Y1 + (1 - T) * Y0   # Observed outcome

    # Convert to PyTorch tensors with explicit dtype
    X_tensor = torch.tensor(X, dtype=torch.float32)
    T_tensor = torch.tensor(T, dtype=torch.float32).reshape
        (-1, 1)
    Y_tensor = torch.tensor(Y, dtype=torch.float32).reshape
        (-1, 1)
    Y0_tensor = torch.tensor(Y0, dtype=torch.float32).
        reshape(-1, 1)
    Y1_tensor = torch.tensor(Y1, dtype=torch.float32).
        reshape(-1, 1)


    return X_tensor, T_tensor, Y_tensor, Y0_tensor,
        Y1_tensor, propensity_scores


# --- 2. DragonNet Architecture (Simplified) ---
class DragonNet(nn.Module):
    def __init__(self, input_dim, hidden_dim=100):
        super(DragonNet, self).__init__()
        self.shared_representation = nn.Sequential(
            nn.Linear(input_dim, hidden_dim),
            nn.ReLU(),
            nn.Linear(hidden_dim, hidden_dim),
            nn.ReLU()
        )
        self.outcome_T0 = nn.Linear(hidden_dim, 1)
        self.outcome_T1 = nn.Linear(hidden_dim, 1)
        self.propensity_head = nn.Sequential(
            nn.Linear(hidden_dim, 1),
            nn.Sigmoid()
        )

    def forward(self, x):
        shared = self.shared_representation(x)
        y0_hat = self.outcome_T0(shared)
```

```python
        y1_hat = self.outcome_T1(shared)
        propensity_hat = self.propensity_head(shared)
        return y0_hat, y1_hat, propensity_hat

    def calculate_targeted_regularization(self, y0_hat,
        y1_hat, t, y):
        """Simplified targeted regularization (example)."""
        # Ensure the output is float32
        return torch.mean((y1_hat - y0_hat) * (t - 0.5)).to(
            torch.float32)


# --- 3. Training Function ---
def train_dragonnet(model, X_train, T_train, Y_train,
    optimizer, epochs=100, batch_size=32,
                    beta=1.0, lambda_reg=0.1):
    """Trains DragonNet."""

    # Ensure input tensors to DataLoader are float32
    X_train = X_train.to(torch.float32)
    T_train = T_train.to(torch.float32)
    Y_train = Y_train.to(torch.float32)

    dataset = TensorDataset(X_train, T_train, Y_train)
    dataloader = DataLoader(dataset, batch_size=batch_size,
        shuffle=True)
    criterion_outcome = nn.MSELoss()
    criterion_propensity = nn.BCELoss()
    model.train()

    for epoch in range(epochs):
        for x_batch, t_batch, y_batch in dataloader:
            optimizer.zero_grad()
            # Ensure inputs to the model are float32
            x_batch = x_batch.to(torch.float32)
            t_batch = t_batch.to(torch.float32)
            y_batch = y_batch.to(torch.float32)

            y0_hat, y1_hat, propensity_hat = model(x_batch)
            y_pred = t_batch * y1_hat + (1 - t_batch) *
                y0_hat

            outcome_loss = criterion_outcome(y_pred, y_batch
                )
            propensity_loss = criterion_propensity(
                propensity_hat, t_batch)
            # Ensure targeted_reg is float32
            targeted_reg = model.
                calculate_targeted_regularization(y0_hat.to(
                torch.float32), y1_hat.to(torch.float32),
                t_batch.to(torch.float32), y_batch.to(torch.
                float32))
```

```python
        # Ensure all components of the loss are float32
            before summing
        loss = outcome_loss.to(torch.float32) + beta *
            propensity_loss.to(torch.float32) +
            lambda_reg * targeted_reg.to(torch.float32)
        loss.backward()
        optimizer.step()
    if (epoch + 1) % 10 == 0:
        print(f'Epoch {epoch + 1}, Loss: {loss.item()}')


# --- 4. Evaluation Function ---
def evaluate_dragonnet(model, X_test, T_test, Y_test,
    Y0_test, Y1_test, propensity_scores_true):
    """Evaluates the trained DragonNet model."""

    model.eval()
    with torch.no_grad():
        # Ensure test data is also float32
        X_test = X_test.to(torch.float32)
        T_test = T_test.to(torch.float32)
        Y_test = Y_test.to(torch.float32)
        Y0_test = Y0_test.to(torch.float32)
        Y1_test = Y1_test.to(torch.float32)

        y0_pred_tensor, y1_pred_tensor,
            propensity_hat_tensor = model(X_test) # Get all
            outputs
        y_pred = (T_test * y1_pred_tensor + (1 - T_test) *
            y0_pred_tensor).squeeze().numpy()
        y0_pred = y0_pred_tensor.squeeze().numpy()
        y1_pred = y1_pred_tensor.squeeze().numpy()
        propensity_hat = propensity_hat_tensor.squeeze().
            numpy()
        Y_test = Y_test.squeeze().numpy()
        Y0_test = Y0_test.squeeze().numpy()
        Y1_test = Y1_test.squeeze().numpy()

        mse = mean_squared_error(Y_test, y_pred)
        ate_pred = np.mean(y1_pred - y0_pred)
        ate_true = np.mean(Y1_test - Y0_test)
        ate_error = np.abs(ate_pred - ate_true)
        pehe = calculate_pehe(Y1_test, Y0_test, y1_pred,
            y0_pred)
        # Convert T_test to NumPy before casting to int
        auc = roc_auc_score(T_test.cpu().squeeze().numpy().
            astype(int), propensity_hat) # Cast T_test to
            int for roc_auc_score

        print("\n--- Evaluation ---")
        print(f"MSE: {mse:.4f}")
        print(f"Predicted ATE: {ate_pred:.4f}")
        print(f"True ATE: {ate_true:.4f}")
```

```python
            print(f"ATE Error: {ate_error:.4f}")
            print(f"PEHE: {pehe:.4f}")
            print(f"Propensity AUC: {auc:.4f}")  # Evaluate
                propensity prediction


def calculate_pehe(y1_true, y0_true, y1_pred, y0_pred):
    """Calculates the Precision in Estimation of
        Heterogeneous Effects (PEHE)."""
    ite_true = y1_true - y0_true
    ite_pred = y1_pred - y0_pred
    return np.sqrt(np.mean((ite_true - ite_pred) ** 2))


# --- 5. Main Execution ---
if __name__=="__main__":
    np.random.seed(42)

    # 5.1. Generate Data
    X, T, Y, Y0, Y1, propensity_scores_true =
        simulate_treatment_effect_data()
    X_train, X_test, T_train, T_test, Y_train, Y_test,
        Y0_train, Y0_test, Y1_train, Y1_test =
        train_test_split(
        X, T, Y, Y0, Y1, test_size=0.2, random_state=42
    )

    # 5.2. Train DragonNet
    dragonnet = DragonNet(input_dim=X.shape[1])
    optimizer_dragonnet = optim.Adam(dragonnet.parameters(),
        lr=0.01)
    train_dragonnet(dragonnet, X_train, T_train, Y_train,
        optimizer_dragonnet)

    # 5.3. Evaluate DragonNet
    # Pass propensity_scores_true to the evaluate function
        if you want to compare
    # predicted propensity scores against the true ones
        during evaluation.
    # However, the evaluate_dragonnet function uses the
        predicted propensity_hat.
    # The original code passed it to evaluate_dragonnet but
        it wasn't used,
    # which is fine. Let's keep the signature consistent
        with the original call.
    evaluate_dragonnet(dragonnet, X_test, T_test, Y_test,
        Y0_test, Y1_test, propensity_scores_true)


    # 5.4. Explanation
    print("\n--- Explanation ---")
    print("This code implements a simplified DragonNet
        architecture for causal inference.")
```

```
print("DragonNet learns a shared representation and has
    heads for:")
print("1.  Outcome prediction (Y0_hat, Y1_hat).")
print("2.  Propensity score estimation (propensity_hat).
    ")
print("The loss function combines outcome loss,
    propensity loss, and a simplified targeted
    regularization term.")
print("Evaluation includes MSE, ATE error, PEHE, and AUC
    for propensity prediction.")
```

8   Exercises

1. What role does the propensity score play in causal inference?
2. How does DragonNet incorporate the propensity score into its architecture?
3. Explain targeted regularization and its purpose in DragonNet.
4. What are the challenges of balancing prediction and propensity loss during training?

References

1. Pearl, Judea. 2009. *Causality: Models, reasoning, and inference*, 2nd. Cambridge University Press.
2. Johansson, Fredrik D., U. Shalit, and D. Sontag. 2016. Learning representations for counterfactual inference. In *Proceedings of the 33rd international conference on machine learning (ICML)*, 1386–1395.
3. Rosenbaum, Paul R., and D. B. Rubin. 1983. The central role of the propensity score in observational studies for causal effects. *Biometrika*, 70.1: 41–55.
4. Goodfellow, Ian, Y. Bengio, and A. Courville. 2016. *Deep learning*. MIT Press.
5. Shi, Chao, D. M. Blei, and V. Veitch. 2019. Adapting neural networks for the estimation of treatment effects. In *Advances in neural information processing systems (NeurIPS)*.
6. Shalit, Uri, F. D. Johansson, and D. Sontag. 2017. Estimating individual treatment effect: Generalization bounds and algorithms. In *Proceedings of the 34th international conference on machine learning (ICML)*, 3076–3085.
7. Hinton, Geoffrey E. 2007. Learning multiple layers of representation. In *Trends in cognitive sciences*, vol. 11.10, 428–434.

Evaluating Causal Models Without Counterfactuals

1 Introduction

In supervised machine learning, evaluating a model is usually straightforward. Given a labeled data set, we can compute standard metrics such as accuracy, mean squared error, or cross-entropy loss by comparing predictions to true labels.

In causal inference [1], however, evaluation becomes much more challenging. This is because for any individual, we can observe only one of the two potential outcomes—the outcome under the treatment they received. The other outcome, the **counterfactual**, is inherently unobservable.

This limitation is known as the **Fundamental Problem of Causal Inference**.

Because we cannot directly observe the true individual treatment effects in real-world data, assessing the accuracy of a causal model's estimates is impossible. Instead, we must rely on indirect methods and creative evaluation strategies.

This chapter will explore different approaches to evaluating causal inference models when counterfactuals are missing. We will introduce proxy metrics, discuss the use of simulated data, and examine real-world evaluation techniques.

Evaluation Challenges in Causal ML

- **The Missing Ground Truth**: In standard machine learning, we have "true labels" to compare our predictions to. In causal inference, the unobserved counterfactuals are like missing labels, making direct evaluation difficult.
- **Beyond Predictive Accuracy**: High predictive accuracy in a machine learning model doesn't guarantee it's learning causal relationships. We need specialized metrics to assess causal validity.

D. Rajamanickam, *Causal Inference for Machine Learning Engineers*,
https://doi.org/10.1007/978-3-031-99680-1_13

- **Evaluation as a Causal Question**: Evaluating a causal model is a causal question: "Does using this model lead to better decisions?" This requires careful consideration of the context of the decision-making.

2 The Core Challenge of Causal Inference

Let us formally restate the central difficulty.

For each individual i, there are two potential outcomes:

- $Y_i(1)$: the outcome if the individual receives treatment,
- $Y_i(0)$: the outcome if the individual does not receive treatment.

However, we can observe only one of these outcomes for any individual. If the individual is treated, we observe $Y_i(1)$; if not treated, we observe $Y_i(0)$.

We cannot simultaneously observe both outcomes for the same individual. This missing data problem makes direct evaluation of causal effect predictions impossible.

In formal terms, if $\hat{\tau}(X)$ is our model's estimate of the Individual Treatment Effect (ITE), we cannot compute the actual error:

$$\hat{\tau}(X) - (Y(1) - Y(0))$$

because $Y(1) - Y(0)$ is not fully observed.

Thus, we must find alternative ways to evaluate models.

The Counterfactual Dilemma

- **One Reality, One Outcome**: The core problem is that we only observe one outcome for each individual—the one that happened. The alternative outcome (counterfactual) is never seen.
- **Individual vs. Average Effects**: While we can estimate average treatment effects across groups, evaluating the accuracy of individual-level causal predictions is inherently limited.
- **Analogy: The Fork in the Road**: It's like judging which path was faster when you can only ever walk one path at a time.

3 Proxy Metrics for Causal Model Evaluation

Although we cannot observe actual individual treatment effects directly, researchers have proposed several proxy metrics that can help assess the quality of causal models.

3.1 Precision in Estimation of Heterogeneous Effects (PEHE)

One important metric is the **Precision in Estimation of Heterogeneous Effects (PEHE) (The Precision in Estimation of Heterogeneous Effects (PEHE) metric** [2] **quantifies how accurately a model estimates individual-level treatment effects** [2].

When both potential outcomes are known (for example, in simulated datasets), PEHE is defined as:

$$\text{PEHE} = \sqrt{\mathbb{E}\left[(\hat{\tau}(X) - \tau(X))^2\right]}$$

where:

- $\hat{\tau}(X)$ is the model's estimated treatment effect,
- $\tau(X) = Y(1) - Y(0)$ is the true treatment effect.

PEHE measures how accurately the model captures individual-level differences in treatment effects.

Lower PEHE values indicate better individualized causal effect estimation.

3.2 Average Treatment Effect (ATE) Error

Another important quantity is the **Average Treatment Effect (ATE)**, defined as:

$$\text{ATE} = \mathbb{E}[Y(1) - Y(0)]$$

If the model estimates $\hat{\tau}(X)$ across the population, the estimated ATE is:

$$\hat{\text{ATE}} = \mathbb{E}[\hat{\tau}(X)]$$

The ATE Error is then:

$$\text{ATE Error} = \left| \hat{\text{ATE}} - \text{ATE} \right|$$

We can compute this error directly in simulated datasets where the true ATE is known. It tells us how accurately the model estimates the overall population effect.

3.3 Policy Risk

Sometimes, the goal of causal modeling is to make treatment decisions, such as deciding which patients should receive a new drug based on predicted benefits.

In such cases, **policy risk** becomes a relevant metric. Policy risk measures the expected loss incurred by following the model's recommendations compared to the optimal policy.

Lower policy risk indicates better decision-making performance.

Policy evaluation often uses surrogate models or assumptions about costs and benefits to approximate the loss.

Proxy Metrics for Evaluation

- **PEHE: Individual-Level Proxy**: PEHE is a proxy for how well the model estimates the treatment effect for *each individual*. It's like measuring the model's ability to personalize its predictions.
- **ATE Error: Population-Level Proxy**: ATE Error measures how well the model estimates the **average** treatment effect across the entire population. It's like measuring the model's overall effectiveness.
- **Policy Risk: Decision-Making Proxy**: Policy Risk evaluates the quality of decisions made **based** on the model's causal estimates. It's like measuring the real-world impact of the model's recommendations.
- **Simulation Dependence**: These proxy metrics often rely on simulated data where we **do** know the counterfactuals, highlighting the importance of simulation in causal machine learning.

4 Simulated Data for Evaluation

Because of the challenges in real-world evaluation, simulated datasets play a vital role in causal machine learning research.

In simulated datasets, we control the data-generating process. Thus, for each individual, we know both $Y(0)$ and $Y(1)$. This allows us to compute exactly PEHE, ATE Error, and other metrics.

Commonly used datasets include:

- **IHDP (Infant Health and Development Program)**: Semi-simulated with real covariates and synthetic outcomes.
- **ACIC Benchmarks**: Fully simulated datasets designed to test causal models under various conditions.
- **Twins Dataset**: Real-world twin birth records where one twin acts as a proxy counterfactual for the other.

Simulated data provides a valuable testing ground for new models. However, models that perform well on simulated data may not always generalize to real-world settings.

Simulation: The Causal Testbed

- **Ground Truth Control**: Simulation allows us to create artificial datasets where we **know** the true causal effects, providing a controlled environment for evaluating causal models.
- **Diverse Scenarios**: We can simulate different types of confounding, noise, and treatment effects to test the model's capabilities thoroughly.
- **Analogy: The Wind Tunnel**: Simulation is like a wind tunnel for testing airplanes; we can control the conditions and measure performance accurately.

5 Real-World Evaluation Strategies

In real-world datasets, where counterfactuals are missing, alternative strategies are needed to evaluate causal models.

5.1 *Natural Experiments and Randomized Controlled Trials (RCTs)*

Sometimes, we can find real-world scenarios where treatment assignment is approximately random due to a natural experiment or a randomized controlled trial (RCT).

In such cases, a simple comparison between treated and control groups can provide unbiased estimates of causal effects, serving as a ground truth for evaluation.

5.2 *Overlap Subsets*

Another strategy is to identify regions of the data where treated and control groups are well-matched in covariates (i.e., there is good overlap). In these regions, simple comparisons may well approximate causal effects.

Causal models can be evaluated based on how well they predict observed outcomes in these subsets.

5.3 *Proxy Outcomes and Surrogate Endpoints*

Sometimes, proxy outcomes strongly correlated with the true causal outcomes are used for evaluation.

For example, in healthcare, surrogate biomarkers can evaluate treatment effects before the final health outcome (such as survival) is observed.

These proxy measures can provide indirect evidence of model quality.

Real-World Causal Evaluation: Imperfect but Necessary

- **Quasi-Experiments**: Natural experiments offer "almost" randomized settings, providing some ground truth for evaluation, similar to how quasi-experimental designs are used in social sciences.
- **Data Subsetting**: Overlap subsets focus on regions where treated and control groups are similar, creating"local" evaluation areas, analogous to techniques used to reduce bias in machine learning.
- **Indirect Measures**: Proxy outcomes use related metrics to indirectly assess causal effects, similar to using proxy metrics for evaluating the performance of recommender systems or language models.

6   Sensitivity Analysis

Since no evaluation method is perfect, performing **sensitivity analysis** and **robustness checks** is essential.

Sensitivity analysis investigates how model conclusions change under different assumptions, such as the presence of unobserved confounding.

Robustness checks test model stability by varying hyperparameters, reweighting samples, or using alternative specifications.

Performing these checks increases confidence in the reliability of causal models in real-world deployment.

Robustness: The Hallmarks of Trustworthy Causal Models

- **Assumption Scrutiny**: Sensitivity analysis tests how much our conclusions depend on the initial assumptions, similar to testing the robustness of a machine learning model to changes in hyperparameters or training data.
- **Model Stability**: Robustness checks assess the model's stability and consistency, ensuring that small changes in the data or model specification don't drastically alter the results.
- **Confidence Building**: These techniques help build confidence in the reliability and trustworthiness of causal models, especially in high-stakes applications.

7 Summary

Evaluating causal inference [1] models presents unique challenges due to the Central Problem of Causal Inference: the inability to observe counterfactual outcomes.

Despite this, researchers have developed proxy metrics such as PEHE, ATE Error, and policy risk to assess model quality, especially on simulated datasets where true effects are known.

In real-world settings, evaluation relies on natural experiments, overlap subsets, proxy outcomes, and sensitivity analyses to build confidence in model validity.

Understanding these evaluation strategies is crucial for interpreting causal model results and ensuring that models are robust and reliable for practical decision-making.

In the next chapter, we will delve into more advanced topics in causal inference, including learning causal graphs from data, dealing with unobserved confounding, decomposing effects, and assessing the robustness of causal conclusions

8 Code

Listing 13.1 Chapter 13: Focusing on evaluating causal models using simulated data and the PEHE and ATE Error metrics

```python
import numpy as np
import pandas as pd
import matplotlib.pyplot as plt
from sklearn.model_selection import train_test_split
from sklearn.linear_model import LinearRegression  # Example
    model
from sklearn.ensemble import RandomForestRegressor  #
    Another example
from sklearn.metrics import mean_squared_error

# --- 1. Simulate Causal Data ---
def simulate_causal_data(n=1000, input_dim=5, ate_mean=2.0,
    ate_std=1.0):
    """Simulates causal data with heterogeneous treatment
        effects and confounding.

    Args:
        n: Number of samples.
        input_dim: Number of input features.
        ate_mean: Mean of the Average Treatment Effect.
        ate_std: Standard deviation of the treatment effect
            heterogeneity.

    Returns:
        Pandas DataFrame with simulated data.
    """

    X = np.random.normal(0, 1, size=(n, input_dim)).astype(
        np.float32)  # Covariates
```

```python
    true_individual_effect = np.random.normal(ate_mean,
        ate_std, size=n).astype(np.float32)
    true_ate = np.mean(true_individual_effect)

    # Propensity scores (probability of treatment)
    propensity_logits = X[:, 0] - 0.5 * X[:, 1] + 0.2 * X[:,
        2]   # Example: depends on first 3 covariates
    propensity_scores = 1 / (1 + np.exp(-propensity_logits))
    T = np.random.binomial(1, propensity_scores, size=n).
        astype(np.float32)   # Treatment

    Y0 = np.dot(X, np.array([0.5, -0.3, 0.2, 0.1, -0.2]).
        reshape(input_dim, 1)).flatten() + np.random.normal
        (0, 1, n)
    Y1 = Y0 + true_individual_effect + np.random.normal(0,
        0.5, n)   # Heterogeneous effects
    Y = T * Y1 + (1 - T) * Y0   # Observed outcome

    return pd.DataFrame({
        'X0': X[:, 0], 'X1': X[:, 1], 'X2': X[:, 2], 'X3': X
            [:, 3], 'X4': X[:, 4],
        'T': T,
        'Y': Y,
        'Y0': Y0,
        'Y1': Y1,
        'true_effect': true_individual_effect
    }), true_ate


# --- 2. Implement PEHE ---
def calculate_pehe(y1_true, y0_true, y1_pred, y0_pred):
    """Calculates the Precision in Estimation of
        Heterogeneous Effects (PEHE)."""

    ite_true = y1_true - y0_true
    ite_pred = y1_pred - y0_pred
    return np.sqrt(np.mean((ite_true - ite_pred) ** 2))


# --- 3. Implement ATE Error ---
def calculate_ate_error(true_ate, predicted_ate):
    """Calculates the error in estimating the Average
        Treatment Effect (ATE)."""

    return np.abs(true_ate - predicted_ate)


# --- 4. Model Training and Prediction (Example) ---
def train_and_predict(model, X_train, Y_train, X_test):
    """Trains a model and makes predictions. (Illustrative)
        """

    model.fit(X_train, Y_train)
```

```python
        Y_pred = model.predict(X_test)
        return Y_pred


# --- 5. Main Execution ---
if __name__=="__main__":
    np.random.seed(42)

    # 5.1. Generate Data
    data, true_ate = simulate_causal_data()

    # 5.2. Split Data
    X = data[['X0', 'X1', 'X2', 'X3', 'X4']].values
    Y = data['Y'].values
    T = data['T'].values
    Y0_true = data['Y0'].values
    Y1_true = data['Y1'].values
    true_effect = data['true_effect'].values

    X_train, X_test, Y_train, Y_test, T_train, T_test,
        Y0_train, Y0_test, Y1_train, Y1_test,
        true_effect_train, true_effect_test =
        train_test_split(
        X, Y, T, Y0_true, Y1_true, true_effect, test_size
            =0.2, random_state=42
    )

# 5.3. Example: Linear Regression Model
model_linear = LinearRegression()
Y_pred_linear = train_and_predict(model_linear, X_train,
    Y_train, X_test)
# It's more appropriate to train separate models for Y0 and
    Y1 on the training data,
# using only the relevant subsets (T_train == 0 for Y0,
    T_train == 1 for Y1) if
# you were trying to replicate potential outcome estimation
    methods.
# However, for this illustrative example just predicting
    Y0_train and Y1_train
# directly on X_test as if they were observed is acceptable
    to demonstrate the metric calculation.
Y0_pred_linear = train_and_predict(model_linear, X_train,
    Y0_train, X_test)
Y1_pred_linear = train_and_predict(model_linear, X_train,
    Y1_train, X_test)

# Pass the true potential outcomes from the test set to
    calculate_pehe
pehe_linear = calculate_pehe(Y1_test, Y0_test,
    Y1_pred_linear, Y0_pred_linear)
ate_pred_linear = np.mean(Y1_pred_linear - Y0_pred_linear)
# Calculate the true ATE on the test set for direct
    comparison
```

```python
ate_true_test = np.mean(true_effect_test)
ate_error_linear = calculate_ate_error(ate_true_test,
    ate_pred_linear) # Use test set ATE
mse_linear = mean_squared_error(Y_test, Y_pred_linear)

print("\n--- Linear Regression Evaluation ---")
print(f"PEHE: {pehe_linear:.4f}")
print(f"ATE Error: {ate_error_linear:.4f}")
print(f"MSE: {mse_linear:.4f}")

# 5.4. Example: Random Forest Model
model_rf = RandomForestRegressor(random_state=42)
Y_pred_rf = train_and_predict(model_rf, X_train, Y_train,
    X_test)
# Similar to the linear model, training on Y0_train and
    Y1_train directly
Y0_pred_rf = train_and_predict(model_rf, X_train, Y0_train,
    X_test) # Changed from Y0_true
Y1_pred_rf = train_and_predict(model_rf, X_train, Y1_train,
    X_test) # Changed from Y1_true

# Pass the true potential outcomes from the test set
pehe_rf = calculate_pehe(Y1_test, Y0_test, Y1_pred_rf,
    Y0_pred_rf) # Changed from Y1_true, Y0_true
ate_pred_rf = np.mean(Y1_pred_rf - Y0_pred_rf)
ate_error_rf = calculate_ate_error(ate_true_test,
    ate_pred_rf) # Use test set ATE
mse_rf = mean_squared_error(Y_test, Y_pred_rf)

print("\n--- Random Forest Evaluation ---")
print(f"PEHE: {pehe_rf:.4f}")
print(f"ATE Error: {ate_error_rf:.4f}")
print(f"MSE: {mse_rf:.4f}")

# 5.5. Visualization (Example)
plt.figure(figsize=(8, 6))
# Visualize true potential outcomes from the test set
plt.scatter(Y0_test, Y1_test, alpha=0.5, label="True
    Potential Outcomes (Test)") # Changed label
plt.scatter(Y0_pred_linear, Y1_pred_linear, alpha=0.5, label
    ="Linear Pred")
plt.scatter(Y0_pred_rf, Y1_pred_rf, alpha=0.5, label="RF
    Pred")
plt.xlabel("Y0 (Outcome if Untreated)")
plt.ylabel("Y1 (Outcome if Treated)")
plt.title("Potential Outcomes: True (Test Set) vs. Predicted
    ") # Changed title
plt.plot([min(Y0_test), max(Y0_test)], [min(Y0_test), max(
    Y0_test)], color='red', linestyle='--', label="Line of
    Equality") # Use test set for limits
plt.legend()
plt.show()
```

9 Exercises

1. Why is evaluating causal models more difficult than evaluating standard supervised models?
2. Define PEHE, ATE Error, and Policy Risk.
3. How can simulated datasets help evaluate causal models?
4. Describe two strategies for evaluating causal models when counterfactuals are unobserved.

References

1. Pearl, Judea. 2009. *Causality: Models, reasoning, and inference*, 2nd. Cambridge University Press.
2. Hill, Jennifer. 2011. Bayesian nonparametric modeling for causal inference. *Journal of Computational and Graphical Statistics* 20.1: 217–240.

Advanced Topics in Causal Inference

1 Introduction

In previous chapters, we assumed that the structure of the causal system was known and that unconfoundedness could be achieved through careful adjustment. However, real-world problems are often more complex. We may not know the causal structure, unmeasured confounders may exist, and we may want to understand whether an effect exists and how it operates.

This chapter explores several advanced topics in causal inference [1]:

- Learning causal graphs from data (causal discovery) [2],
- Dealing with unobserved confounding using instrumental variables [3],
- Decomposing effects into direct and indirect components (mediation analysis),
- Assessing the robustness of causal conclusions through sensitivity analysis.

Advanced Causal Methods for ML

- **Beyond Basic Adjustment**: This chapter moves beyond basic confounding adjustment to address more complex causal scenarios relevant to machine learning.
- **Learning Causal Structure**: We explore methods to learn the causal relationships directly from data, which could automate feature engineering and model design aspects.
- **Robustness and Generalization**: We discuss techniques to improve the robustness and generalization of machine learning models by considering causal mechanisms.

© The Author(s), under exclusive license to Springer Nature Switzerland AG 2025
D. Rajamanickam, *Causal Inference for Machine Learning Engineers*,
https://doi.org/10.1007/978-3-031-99680-1_14

2   Causal Discovery

2.1   Motivation for Causal Discovery

Often, we do not know the causal relationships among variables beforehand. Causal discovery [2] attempts to learn the causal graph directly from observational or interventional data.

However, it is difficult because many graphs can generate the same observed statistical dependencies. Therefore, causal discovery [2] requires strong assumptions such as:

- **Causal sufficiency**: No unobserved common causes,
- **Faithfulness**: All observed independencies reflect causal separations.

2.2   The PC Algorithm

The **PC Algorithm** [4] is a well-known method for causal discovery.
It proceeds in two main steps:

1. **Skeleton Identification**: Start with a fully connected, undirected graph. Remove edges between pairs of variables that are conditionally independent given some set of other variables, based on statistical tests.
2. **Orientation of Edges**: Apply rules to direct edges based on detected collider structures (where two arrows converge at a node) and logical propagation to preserve acyclicity (Fig. 1).

The output is a **Partially Directed Acyclic Graph** (PDAG) where some causal directions may remain undetermined.

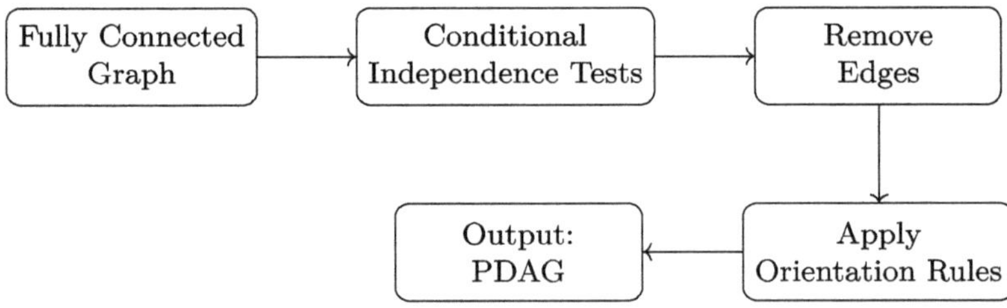

Fig. 1   Flow of the PC algorithm for causal discovery

2.3 *Greedy Equivalence Search (GES)*

Greedy Equivalence Search (GES) is a score-based method. It searches over the space of possible graphs to maximize a score like the Bayesian Information Criterion (BIC).

It operates in two phases:

- **Forward Phase**: Add edges that most improve the score.
- **Backward Phase**: Remove edges that least decrease the score.

GES can be more robust to noise than constraint-based methods like PC but still assumes causal sufficiency.

2.4 *LiNGAM Model*

The **Linear Non-Gaussian Acyclic Model (LiNGAM)** offers another approach based on distributional assumptions.

LiNGAM assumes:

- The data-generating process is linear,
- The noise terms are non-Gaussian and mutually independent.

The model is:

$$X = BX + e$$

where:

- X is a vector of observed variables,
- B is a matrix of causal coefficients with zeros on and above the diagonal (no cycles),
- e is a vector of independent, non-Gaussian noise terms.

Non-Gaussianity allows LiNGAM to uniquely recover the causal order, unlike methods that rely only on second moments (like correlation).

Causal Discovery: Learning the Causal Graph

- **Automated Feature Engineering**: Causal discovery [2] algorithms can automate the process of feature engineering by identifying relevant causal features and relationships directly from data.
- **Graph-Based Representations**: The learned causal graph can be used as a structural prior for graph neural networks [5] or other machine learning models that operate on graph-structured data.

- **Limitations and Assumptions**: Acknowledging the strong assumptions (e.g., causal sufficiency, faithfulness) required for causal discovery [2], which may not always hold in real-world machine learning applications.

3   Instrumental Variables (IV)

3.1   *Motivation*

When unmeasured confounding exists, standard adjustment techniques fail. **Instrumental variables (IVs)** help recover causal effects under these conditions.

    An IV affects the treatment but influences the outcome only through the treatment and is independent of unobserved confounders.

3.2   *Causal Diagram for IV*

3.3   *IV Assumptions*

For a valid IV:

- **Relevance**: Z must affect T.
- **Exclusion Restriction**: Z affects Y only through T.
- **Independence**: Z is independent of unmeasured confounders U.

3.4   *Two-Stage Least Squares (2SLS)*

The most popular method to estimate causal effects using IVs is **Two-Stage Least Squares (2SLS)** (Fig. 2).

1. **First Stage**: Regress T on Z to predict $\hat{T}$:

$$T = \pi_0 + \pi_1 Z + \varepsilon$$

    where $\hat{T}$ captures the variation in T due to Z.

2. **Second Stage**: Regress Y on $\hat{T}$:

$$Y = \beta_0 + \beta_1 \hat{T} + \eta$$

The coefficient β_1 estimates the causal effect of T on Y.

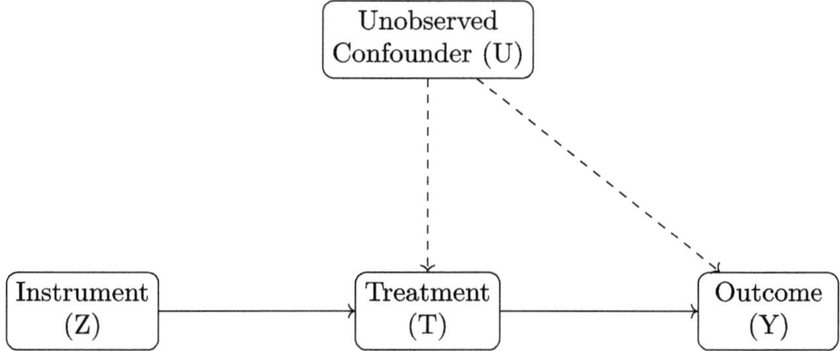

Fig. 2 Causal diagram illustrating an instrumental variable

Key intuition: The first stage isolates the "clean" part of treatment variation, and the second stage uses it to estimate causal impact.

Instrumental Variables: Dealing with Unobserved Confounding

- **Hidden Confounders in ML**: Unobserved confounders are analogous to hidden biases or latent variables that can distort the relationships learned by machine learning models.
- **IVs as Debiasing Tools**: Instrumental variables [3] offer a way to debias machine learning models by isolating the causal effect of a feature even when unobserved confounders are present.
- **Analogy: The Clean Experiment**: An instrumental variable creates a "cleaner" experiment within the observational data, allowing the model to learn a more accurate causal relationship.

4 Mediation Analysis

4.1 Motivation

In some cases, we are interested in how a treatment affects an outcome through an intermediate variable, called a **mediator**.

4.2 Causal Diagram with a Mediator

See Fig. 3.

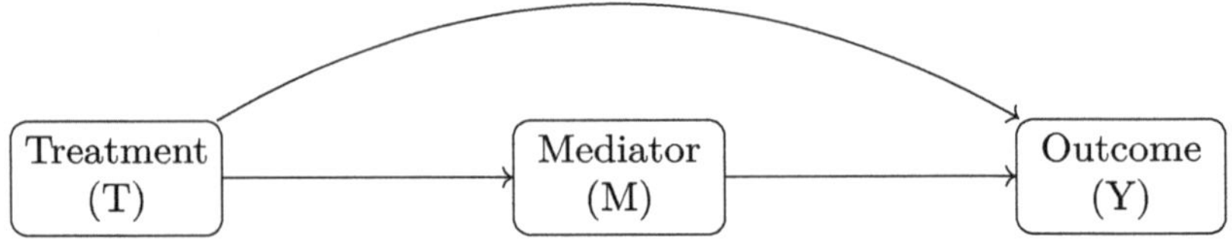

Fig. 3 Causal diagram illustrating mediation

4.3 Definitions: Direct and Indirect Effects

The total effect of treatment on outcome can be decomposed into:

- **Natural Direct Effect (NDE):** The part of the effect not passing through the mediator.
- **Natural Indirect Effect (NIE):** The part of the effect passing through the mediator.

  Formally:
$$\text{Total Effect} = \text{NDE} + \text{NIE}$$

4.4 Mediation Formula

The decomposition is expressed as:

$$\mathbb{E}[Y(T = 1, M(T = 1))] - \mathbb{E}[Y(T = 0, M(T = 0))] = \text{NDE} + \text{NIE}$$

Here, $Y(T, M)$ denotes the outcome under specific treatment and mediator values.

4.5 Estimating Mediation Effects

We typically:

- Model the mediator as a function of treatment.
- Model the outcome as a function of treatment and mediator.

Counterfactual outcomes can be simulated using these models to estimate direct and indirect effects.

Mediation Analysis: Understanding Causal Mechanisms

- **Causal Pathways in ML:** Mediation analysis helps to understand the causal pathways within a machine learning model, identifying how input features influence the outcome through intermediate representations or processing steps.

- **Model Interpretability**: This can improve the interpretability of complex models like neural networks by revealing the mechanisms through which they arrive at their predictions.
- **Analogy: The Flow of Influence**: Mediation analysis is like tracing the influence flow through a network to understand how each node contributes to the final output.

5 Sensitivity Analysis

5.1 Motivation

Causal conclusions are only as firm as the assumptions behind them. **Sensitivity analysis** assesses how robust conclusions are to possible violations, like unmeasured confounding.

5.2 Rosenbaum Bounds

Rosenbaum bounds model the effect of hidden bias:

$$\frac{1}{\Gamma} \le \frac{P(T = 1 \mid X, U)}{P(T = 1 \mid X, U')} \le \Gamma$$

where Γ measures how much hidden bias would need to exist to overturn the result.

- $\Gamma = 1$ indicates no hidden bias.
- Higher Γ values imply more allowed bias.

5.3 Interpretation

If the conclusion remains stable even under large Γ, the result is robust. The result is sensitive if small Γ overturns the finding.

Sensitivity Analysis: Assessing Robustness to Assumptions

- **Assumption Validation in ML**: Sensitivity analysis is analogous to robustness checks in machine learning, where we evaluate how sensitive the model's predictions are to data or specification changes.
- **Uncertainty Quantification**: It helps to quantify the uncertainty in our causal estimates due to potential violations of assumptions, providing a more cautious and reliable interpretation of the results.

- **Model Criticism**: Sensitivity analysis can be used to criticize and improve machine learning models by identifying their weaknesses and limitations.

6 Summary

In this chapter, we introduced important advanced concepts:

- Learning causal graphs from data,
- Using instruments to identify effects under unmeasured confounding,
- Decomposing total effects into direct and mediated pathways,
- Assessing robustness through sensitivity analysis.

These techniques open new possibilities for richer, more reliable causal inference in complex settings.

7 Code

Listing 14.1 Chapter 14: covering Causal Discovery (simplified PC Algorithm) and Instrumental Variables (2SLS)

```
import numpy as np
import pandas as pd
import statsmodels.formula.api as sm
import networkx as nx
import matplotlib.pyplot as plt
from itertools import combinations

# --- 1. Simulate Data for Causal Discovery (Simplified) ---
def simulate_causal_discovery_data(n=1000):
    """Simulates data for a simple causal discovery example
       (A -> B -> C, A -> C)."""

    A = np.random.normal(0, 1, n)
    B = 0.8 * A + np.random.normal(0, 1, n)
    C = 0.5 * A + 0.7 * B + np.random.normal(0, 1, n)
    return pd.DataFrame({'A': A, 'B': B, 'C': C})


# --- 2. Simplified PC Algorithm (Illustrative) ---
def simplified_pc_algorithm(data, alpha=0.05):
    """A highly simplified version of the PC algorithm for
       illustration."""

    variables = list(data.columns)
    graph = nx.Graph()   # Undirected initially
    graph.add_nodes_from(variables)
```

```python
        graph.add_edges_from(combinations(variables, 2))  #
            Start fully connected

        def is_independent(var1, var2, cond_set=None):
            """A very basic independence test (p-value from
                linear regression)."""
            if cond_set:
                formula = f'{var1} ~ ' + ' + '.join(cond_set)
            else:
                formula = f'{var1} ~ 1'  # Intercept only
            model = sm.ols(formula, data=data).fit()
            return model.f_pvalue > alpha  # Simplified: using F
                -test p-value

        # Very simplified: Only 0-variable conditioning
        for var1, var2 in list(graph.edges):  # Iterate over a
            copy to allow removal
            if is_independent(var1, var2):
                graph.remove_edge(var1, var2)

        # Very simplified: Limited orientation (collider
            detection)
        directed_graph = nx.DiGraph(graph)  # Convert to
            directed
        for a, b, c in combinations(variables, 3):
            if graph.has_edge(a, b) and graph.has_edge(b, c) and
                not graph.has_edge(a, c):
                if not is_independent(a, c, cond_set=[b]):
                    if directed_graph.has_edge(b, a):
                        directed_graph.remove_edge(b, a)
                    if directed_graph.has_edge(b, c):
                        directed_graph.remove_edge(b, c)
                    directed_graph.add_edge(a, b)
                    directed_graph.add_edge(c, b)  # Collider at
                        B

    return directed_graph


# --- 3. Simulate Data for Instrumental Variables ---
def simulate_iv_data(n=1000, effect_t_y=2, effect_z_t=0.5,
    effect_u_t=0.3, effect_u_y=0.8):
    """Simulates data for an Instrumental Variables scenario
        (Z -> T -> Y, U -> T, U -> Y)."""

    U = np.random.normal(0, 1, n)  # Unobserved confounder
    Z = np.random.normal(0, 1, n)  # Instrument
    T = 0.5 * Z + 0.3 * U + np.random.normal(0, 1, n)
    Y = effect_t_y * T + 0.8 * U + np.random.normal(0, 1, n)
    return pd.DataFrame({'Z': Z, 'T': T, 'Y': Y, 'U': U})


# --- 4. Implement Two-Stage Least Squares (2SLS) ---
```

```python
def implement_2sls(data):
    """Implements Two-Stage Least Squares (2SLS) to estimate
        the effect of T on Y."""

    # 4.1. First Stage: Regress T on Z
    first_stage = sm.ols('T ~ Z', data=data).fit()
    data['T_hat'] = first_stage.predict(data)  # Predicted T

    # 4.2. Second Stage: Regress Y on T_hat
    second_stage = sm.ols('Y ~ T_hat', data=data).fit()
    effect_of_t = second_stage.params['T_hat']

    print("\n--- Two-Stage Least Squares (2SLS) ---")
    print("Estimated effect of T on Y:", effect_of_t)
    return effect_of_t


# --- 5. Main Execution ---
if __name__=="__main__":
    np.random.seed(42)

    # 5.1. Causal Discovery Example
    causal_data = simulate_causal_discovery_data()
    learned_graph = simplified_pc_algorithm(causal_data)

    plt.figure(figsize=(8, 6))
    nx.draw_circular(learned_graph, with_labels=True,
        node_size=2000, node_color='lightyellow', font_size
        =12, arrowsize=20)
    plt.title('Learned Causal Graph (Simplified PC)')
    plt.show()

    print("\n--- Causal Discovery ---")
    print("This demonstrates a simplified PC algorithm.")
    print("It learns the causal structure from data (A -> B
        -> C, A -> C) by testing conditional independence.")
    print("Note: This is a simplified version; the full PC
        algorithm is more complex.")

    # 5.2. Instrumental Variables Example
    iv_data = simulate_iv_data()
    estimated_effect_iv = implement_2sls(iv_data)
    print("\n--- Instrumental Variables ---")
    print("This demonstrates Two-Stage Least Squares (2SLS).
        ")
    print("It estimates the effect of T on Y using an
        instrument Z to address unobserved confounding.")
```

8   Exercises

1. **Causal Discovery** (a) Explain why purely observational data cannot uniquely determine the true causal graph without additional assumptions. (b) Briefly describe the two main phases of the PC Algorithm [4].

2. **Instrumental Variables** (a) List and explain the three key assumptions for a valid instrumental variable. (b) Suppose you want to estimate the effect of exercise on weight loss but suspect unobserved confounding. Suggest a possible instrumental variable and justify your choice.

3. **Mediation Analysis** (a) Define natural direct effect (NDE) and natural indirect effect (NIE). (b) In a study examining the effect of a new diet (treatment) on blood pressure (outcome) through weight change (mediator), sketch the causal diagram and explain how you would estimate the direct and indirect effects.

4. **Sensitivity Analysis** (a) What does the parameter Γ represent in Rosenbaum bounds sensitivity analysis? (b) If your study shows robustness up to $\Gamma = 2.0$, what does this imply about the strength of unobserved confounding needed to overturn your causal conclusion?

5. **Critical Thinking** Imagine you find a strong causal effect of an online education program on test scores. However, you suspect that motivated students might self-select into the program. (a) Propose how to use an instrumental variable to deal with this problem. (b) Describe how sensitivity analysis could help assess the robustness of your findings if no valid instrument is available.

References

1. Pearl, Judea. 2009. Causality: Models, reasoning, and inference. 2nd ed. Cambridge University Press.
2. Spirtes, Peter, Clark, N. Glymour, and Richard, Scheines. 2000. Causation, prediction, and search. MIT Press.
3. Angrist, Joshua, D., Guido, W. Imbens, and Donald, B. Rubin. 1996. Identification of causal effects using instrumental variables. *Journal of the American Statistical Association* 91 (434): 444–455.
4. Spirtes, Peter, and Clark, Glymour. 1991. An algorithm for fast recovery of sparse causal graphs. *Social Science Computer Review* 9 (1): 62–72.
5. Scarselli, Franco, et al. 2009. The graph neural network model . *IEEE Transactions on Neural Networks* 20 (1): 61–80.

Assumptions and Real-World Challenges in Causal Inference

1 Evaluation of Key Assumptions

Causal inference [1] is primarily reliant on a set of core assumptions. While these assumptions enable identification of causal effects, their validity can vary significantly across domains and data settings. This chapter expands on the assumptions already introduced, focusing on their implications, limitations, and diagnostic strategies.

1.1 Ignorability (Unconfoundedness)

Ignorability assumes that treatment assignment is independent of potential outcomes given observed covariates:

$$Y(1), Y(0) \perp\!\!\!\perp T \mid X$$

Contexts where it may hold:

- Randomized Controlled Trials (RCTs), where randomization breaks the link between treatment and potential outcomes.
- Observational studies with rich covariate sets capturing confounders.

Challenges:

- Unmeasured confounding is often difficult to rule out.
- Diagnostics (e.g., balance checks) can assess observed covariates, not hidden ones.

1.2 Positivity (Overlap)

Positivity requires that every unit has a non-zero probability of receiving each treatment:

$$0 < P(T = 1 \mid X = x) < 1 \quad \text{for all } x$$

Violations in practice:

- Highly stratified populations where treatment is deterministic based on some covariates.
- Automated eligibility rules (e.g., in healthcare or hiring algorithms).

Solutions:

- Trimming or reweighting.
- Restricting the population to regions of common support.

1.3 Stable Unit Treatment Value Assumption (SUTVA)

SUTVA posits no interference and no hidden variation in treatment:

- Each unit's outcome depends only on its treatment.
- The treatment is consistently applied.

SUTVA breaks in:

- Social network contexts (e.g., spillover effects).
- Clustered interventions.

1.4 Correct Model Specification

Many estimators assume a correctly specified functional form. Misspecification can lead to biased estimates even if assumptions like ignorability hold.

Remedies:

- Flexible machine learning models (e.g., Double Machine Learning) [2].
- Cross-validation and model averaging.

Causal Assumptions in ML

- **Assumption Awareness**: Machine learning engineers should know the underlying causal assumptions, as they influence the validity and reliability of the model's predictions and decisions.

- **Assumption Violations as Model Weaknesses**: Violations of causal assumptions can be seen as weaknesses or vulnerabilities in machine learning models, leading to biased or inconsistent results.
- **Connecting to ML Best Practices**: Techniques in machine learning, such as data augmentation, regularization, and model validation, can be interpreted as ways to mitigate the impact of assumption violations.

2 Real-World Data Challenges

2.1 Missing Data

Real datasets often contain missing covariates, treatments, or outcomes. Standard causal estimators may be biased if missingness is not completely random.

Approaches:

- Multiple imputation.
- Inverse probability weighting for missingness.
- Sensitivity analysis.

2.2 Measurement Error

Measurement error in treatment or covariates can attenuate or bias causal effects.

Solutions:

- Instrumental variable methods (if valid instruments are available).
- Error modeling and correction methods.

2.3 Selection Bias and Censoring

In practice, data may be selected based on post-treatment variables (e.g., only observing patients who survived) or censored (e.g., loss to follow-up).

Mitigation:

- Structural causal models [1] to encode selection mechanisms.
- Inverse probability of censoring weights.

2.4 Heterogeneous Treatment Effects

Real-world effects often vary across subpopulations. Average treatment effects may mask critical subgroup patterns.

Approaches:

- Causal forests, meta-learners (T-learner, X-learner, etc.).
- Subgroup analyses and stratification.

2.5 High Dimensionality and Regularization

High-dimensional covariates are standard in modern data (e.g., genomics, EHR). This poses challenges for estimation, interpretability, and inference.

Solutions:

- Regularization (Lasso, Elastic Net).
- Sample-splitting techniques for valid inference.

Real-World Data Issues in ML

- **Data Quality is Crucial**: Machine learning models are susceptible to data quality issues, and causal inference is no exception. Missing data, measurement error, and selection bias can severely impact the accuracy and reliability of causal estimates.
- **Generalization Challenges**: These data challenges can hinder the generalization ability of machine learning models, causing them to perform poorly on new or unseen data.
- **ML Solutions and Causal Counterparts**: Techniques in machine learning, such as imputation, regularization, and ensemble methods, have parallels in causal inference and can be used to address some of these data challenges.

3   Conclusion

Assumptions are the backbone of causal inference. A deep understanding of their roles, limitations, and empirical diagnostics is essential for valid inference. Likewise, applying causal methods to real-world data demands careful handling of noise, biases, and data quality issues. This chapter provides a practical perspective on navigating these challenges to ensure robustness and interpretability in causal analysis.

References

1. Pearl, Judea. 2009. Causality: Models, reasoning, and inference. 2nd ed. Cambridge University Press.
2. Chernozhukov, Victor, et al. 2018. Double/debiased machine learning for treatment and structural parameters . *The Econometrics Journal* 21 (1): C1–C68.

Summary of Key Concepts

1 Looking Back: What We Have Learned

In this book, we embarked on a journey through causal inference [1] and deep learning [2]. We began with the foundational concepts of causal thinking, distinguishing causal relationships from mere correlations. We learned how treatments and outcomes are connected, and how confounding variables can bias naive comparisons.

We then built our understanding of causal graphs, interventions, and counterfactuals, gaining tools like do-calculus [1] to formalize causal reasoning.

Moving forward, we saw how classical causal inference methods estimate treatment effects and why they struggle in high-dimensional, complex settings. This motivated the need for integrating deep learning techniques.

The book's second part explored how deep learning models such as TARNet [3], CFRNet [4], and DragonNet [5] bring powerful representation learning [6] capabilities to causal tasks. These models allow us to handle rich, unstructured data while addressing challenges like confounding and selection bias.

Finally, we tackled the complex problems of evaluating causal models without counterfactuals, understanding proxy metrics such as PEHE, ATE Error, and policy risk, and using simulated and real-world strategies for validation.

Along the way, we learned that while deep learning models offer great flexibility, careful design and a deep respect for causal principles are necessary to avoid pitfalls and produce meaningful conclusions.

2 The Importance of Causal Thinking

At the heart of this journey lies a fundamental lesson: **causal thinking matters**.

In an age where machine learning systems are deployed in critical decision-making domains 'healthcare, education, criminal justice, finance' making accurate

© The Author(s), under exclusive license to Springer Nature Switzerland AG 2025

D. Rajamanickam, *Causal Inference for Machine Learning Engineers*,

https://doi.org/10.1007/978-3-031-99680-1_16

predictions is not enough. We must understand the effects of actions, reason about interventions, and model not just what is but what could be.

Causal inference provides a principled framework for answering questions about cause and effect. Integrating this framework into modern deep learning methods enables us to move beyond prediction toward decision-making systems that are fairer, more interpretable, and more robust.

Causality: The Foundation of Reliable AI

- **Decision-Making in the Real World**: As machine learning models are increasingly used to make decisions in healthcare, finance, and other critical domains, causal thinking becomes essential for ensuring those decisions are sound and reliable.
- **Fairness and Ethical AI**: Causal reasoning provides a principled way to identify and mitigate biases in machine learning models, promoting fairness and ethical AI development.
- **The Future of AI**: The ability to reason causally is crucial to building truly intelligent systems that can understand and interact with the world meaningfully.

3   Emerging Directions in Causal Deep Learning

As the field evolves, several exciting frontiers emerge at the intersection of causality and deep learning. Recent research on causal representation learning [7] aims to disentangle latent variables to reflect the underlying causal structure of the data.

3.1   Causal Representation Learning

One central area of research is **causal representation learning** [7]. The goal here is to learn latent representations that reflect the underlying causal structure of the data, not just compress information for prediction.

Future models may be able to discover disentangled [7] factors of variation that correspond to meaningful causal variables, automatically uncovering causal graphs from high-dimensional inputs like images or text.

3.2   Causal Reinforcement Learning

Causal reinforcement learning is another promising direction. In reinforcement learning, agents learn from interactions with the environment. Incorporating causal reasoning into agents' decision-making can help them learn more efficiently, generalize across environments, and avoid spurious correlations.

For example, agents could use counterfactual reasoning to ask: "What would have happened if I had taken a different action?" and improve their policies accordingly.

3.3 Causality in Large Language Models

As large language models (LLMs) like GPT and BERT become increasingly prevalent, there is growing interest in making these models causally aware.

LLMs are trained primarily for correlation-based tasks, such as predicting the next word based on context. Future work may focus on teaching language models to understand causal relationships in text, reason about interventions, and generate counterfactual scenarios.

Such models could power applications in scientific discovery, policy analysis, and more.

3.4 Fairness, Robustness, and Interpretability

Causal methods offer pathways to improve **fairness**, **robustness**, and **interpretability** in machine learning.

By explicitly modeling causal relationships, we can better detect and correct biases, ensure models are robust to distribution shifts, and provide more understandable explanations for model decisions.

These goals are critical for deploying AI systems responsibly in high-stakes domains.

The Causal Deep Learning Frontier

- **Causal Representation Learning: Uncovering the True Variables**: This aims to learn feature representations that reflect the underlying causal structure of the data, going beyond simple pattern recognition.
- **Causal Reinforcement Learning: Intelligent Action**: Integrating causal reasoning into reinforcement learning allows agents to learn more efficient and robust policies by understanding the consequences of their actions.
- **Causality in Large Language Models: Understanding Meaning**: Teaching large language models to understand causal relationships in text is crucial for natural language understanding and generation.
- **Fairness, Robustness, and Interpretability: Building Trustworthy AI**: Causal methods offer promising avenues for building fairer, more robust, and more interpretable machine learning systems, which is essential for responsible AI deployment.

4   Closing Thoughts

Causal inference and deep learning are two of the most powerful ideas in modern data science. Individually, each field offers remarkable capabilities. Together, they hold the potential to transform the way we build intelligent systems.

By combining the representational power of deep learning with the rigorous reasoning of causal inference, we can build models that predict, understand, fit the data, and guide actions in the real world.

The journey is just beginning. As you continue your exploration of causal deep learning, remember to think carefully about the assumptions you make, the questions you ask, and the goals you pursue.

As in all science, humility, curiosity, and rigor are your greatest allies in causality.

The best way to predict the future is to understand the causes that shape it.

5   Exercises

1. Summarize the main challenges in causal deep learning.
2. What is causal representation learning [7], and why is it important?
3. How might causality improve reinforcement learning agents?
4. Discuss one potential application area where causal deep learning could have a major impact.

References

1. Pearl, Judea. 2009. *Causality: Models, reasoning, and inference*. 2nd ed. Cambridge University Press.
2. Ian Goodfellow, Yoshua Bengio, and Aaron Courville. 2016. *Deep learning*. MIT Press.
3. Shalit, Uri, Fredrik, D. Johansson, and David, Sontag. 2017. *Estimating individual treatment effect: Generalization bounds and algorithms*. In *Proceedings of the 34th international conference on machine learning (ICML)*, 3076–3085.
4. Johansson, Fredrik, D., Uri, Shalit, and David, Sontag. 2016. Learning representations for counterfactual inference. In *Proceedings of the 33rd international conference on machine learning (ICML)*. 1386–1395.
5. Shi, Chao, David, M. Blei, and Victor, Veitch. Adapting neural networks for the estimation of treatment effects. In *Advances in neural information processing systems (NeurIPS)*.
6. Hinton, Geoffrey, E. 2007. Learning multiple layers of representation. *Trends in Cognitive Sciences* 11 (10): 428–434.
7. Schölkopf, Bernhard, et al. 2021. Toward causal representation learning. *Proceedings of the IEEE* 109 (5): 612–634.

Case Studies

1 Case Study 1: Estimating the Effect of an Exercise Program on Blood Pressure

1.1 Problem Description

Suppose we are interested in assessing whether participating in a structured exercise program causally reduces an individual's blood pressure.

However, participation in the program may not be randomly assigned. People who are already health-conscious are more likely to join and may already have lower blood pressure.

Thus, we must adjust for confounding factors such as baseline health and age.

1.2 Causal Graph (DAG)

The causal relationships can be represented as (Fig. 1):

1.3 Simulating Data

We simulate a dataset following this causal structure:

Listing 17.1 Simulating exercise program data

```
import numpy as np

np.random.seed(42)
n = 2000
Age = np.random.normal(50, 10, n)      # Mean age 50
Health = np.random.normal(0, 1, n)     # Baseline health score
```

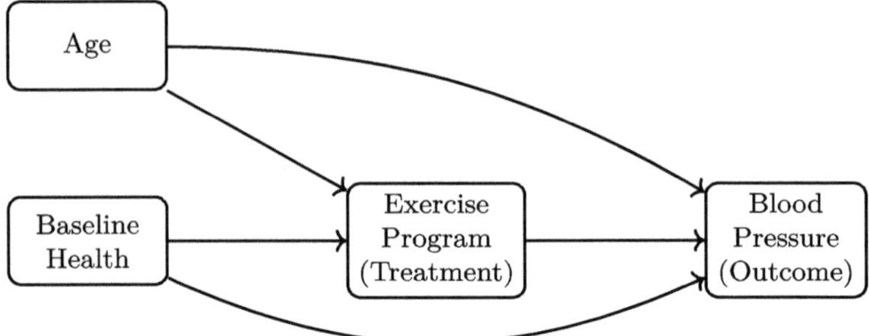

Fig. 1 Causal diagram: effect of an exercise program on blood pressure, with age and baseline health as confounders

```python
# Propensity to participate based on health and age
logit_p = 0.05 * Age + 0.8 * Health
p = 1 / (1 + np.exp(-logit_p))
Exercise = np.random.binomial(1, p)

# Blood pressure outcome
BP_baseline = 140 - 0.3 * Health - 0.2 * Age + np.random.
    normal(0, 5, n)
Treatment_effect = -5  # Exercise reduces BP by 5 units
BloodPressure = BP_baseline + Exercise * Treatment_effect

# Assemble into features
X = np.stack([Age, Health], axis=1)
```

1.4   Modeling with CFRNet

We train a CFRNet model to predict the effect of exercise on blood pressure.

Listing 17.2   Training CFRNet on exercise program data

```python
# Split data into train/test
from sklearn.model_selection import train_test_split
X_train, X_test, T_train, T_test, Y_train, Y_test =
    train_test_split(X, Exercise, BloodPressure, test_size
    =0.3)

# Convert to tensors
import torch
X_train_tensor = torch.tensor(X_train, dtype=torch.float32)
T_train_tensor = torch.tensor(T_train, dtype=torch.float32).
    unsqueeze(1)
Y_train_tensor = torch.tensor(Y_train, dtype=torch.float32).
    unsqueeze(1)
```

```python
# Train CFRNet (reusing earlier CFRNet class)
model = CFRNet(input_dim=2)
optimizer = torch.optim.Adam(model.parameters(), lr=0.01)
criterion = torch.nn.MSELoss()

for epoch in range(100):
    optimizer.zero_grad()
    pred = model(X_train_tensor, T_train_tensor)
    loss = criterion(pred, Y_train_tensor)
    loss.backward()
    optimizer.step()
```

1.5 Evaluation

Using true counterfactuals from the simulation, we evaluate the model.

Listing 17.3 Evaluating CFRNet performance

```python
with torch.no_grad():
    y0_pred = model(torch.tensor(X_test, dtype=torch.float32
        ), torch.zeros((len(X_test),1)))
    y1_pred = model(torch.tensor(X_test, dtype=torch.float32
        ), torch.ones((len(X_test),1)))

pred_effect = (y1_pred - y0_pred).squeeze().numpy()
true_effect = np.full_like(pred_effect, Treatment_effect)

# PEHE
pehe = np.sqrt(np.mean((pred_effect - true_effect)**2))

# ATE Error
true_ate = np.mean(true_effect)
pred_ate = np.mean(pred_effect)
ate_error = np.abs(pred_ate - true_ate)

print(f"PEHE: {pehe:.4f}")
print(f"ATE Error: {ate_error:.4f}")
```

1.6 Interpretation

If the model is well-trained, we expect a low PEHE and a small ATE error, indicating that the CFRNet successfully estimated the causal effect of the exercise program on blood pressure.

Causal Modeling in Healthcare with ML

- **Treatment Effect Prediction**: This case study demonstrates how machine learning models can predict the causal effect of a treatment (exercise program) on a health outcome (blood pressure).
- **Confounding Adjustment with ML**: CFRNet is used to adjust for confounding variables (age, baseline health), a crucial step in obtaining unbiased estimates, similar to how machine learning models perform feature selection to control extraneous variables.
- **Personalized Medicine**: Estimating individual treatment effects, as done implicitly by CFRNet, is a step towards personalized medicine, where treatment plans are tailored to individual patient characteristics.

2　Case Study 2: Estimating the Effect of Online Learning on Student Performance

2.1　*Problem Description*

We are interested in estimating the causal effect of participating in an online learning program on students' academic performance.

However, students self-select into the program based on motivation and prior grades, which introduces confounding. Thus, we must account for confounders like previous GPA and study habits.

2.2　*Causal Graph (DAG)*

The causal structure can be represented as (Fig. 2):

2.3　*Simulating Data*

We simulate a dataset following this causal structure:

Listing 17.4 Simulating online learning data

```python
import numpy as np

np.random.seed(123)
n = 3000
Motivation = np.random.normal(0, 1, n)
GPA = np.random.uniform(2.0, 4.0, n)
```

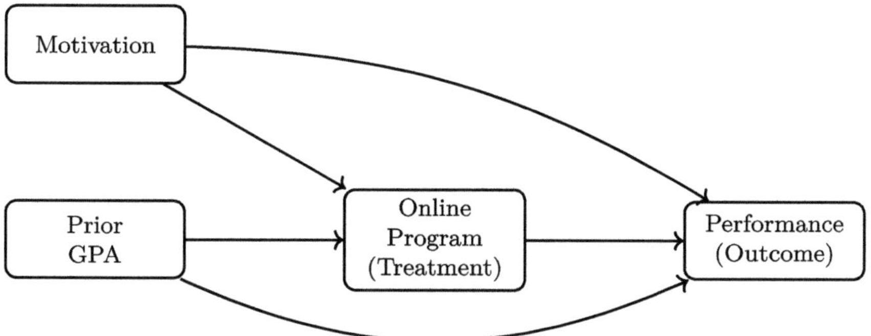

Fig. 2 Causal diagram: online program effect on performance, with prior GPA and motivation as confounders

```python
# Propensity to join online program
logit_p = 0.7 * Motivation + 0.5 * (GPA - 3.0)
p = 1 / (1 + np.exp(-logit_p))
OnlineProgram = np.random.binomial(1, p)

# Academic performance
Performance_baseline = 70 + 5 * Motivation + 10 * (GPA -
    3.0) + np.random.normal(0, 5, n)
Treatment_effect = 5  # Online learning boosts scores by 5
    points
Performance = Performance_baseline + OnlineProgram *
    Treatment_effect

# Assemble into features
X = np.stack([Motivation, GPA], axis=1)
```

2.4 *Modeling with DragonNet*

We train a DragonNet [1] model to predict the effect of online learning.

Listing 17.5 Training DragonNet on online learning data

```python
# Split data
from sklearn.model_selection import train_test_split
X_train, X_test, T_train, T_test, Y_train, Y_test =
    train_test_split(X, OnlineProgram, Performance,
    test_size=0.3)

# Convert to tensors
import torch
X_train_tensor = torch.tensor(X_train, dtype=torch.float32)
T_train_tensor = torch.tensor(T_train, dtype=torch.float32).
    unsqueeze(1)
```

```python
Y_train_tensor = torch.tensor(Y_train, dtype=torch.float32).
    unsqueeze(1)

# Train DragonNet (reusing earlier DragonNet class)
model = DragonNet(input_dim=2)
optimizer = torch.optim.Adam(model.parameters(), lr=0.01)
criterion_outcome = torch.nn.MSELoss()
criterion_propensity = torch.nn.BCELoss()

for epoch in range(100):
    optimizer.zero_grad()
    y_pred, t_pred = model(X_train_tensor, T_train_tensor)
    loss_outcome = criterion_outcome(y_pred, Y_train_tensor)
    loss_propensity = criterion_propensity(t_pred,
        T_train_tensor)
    loss = loss_outcome + loss_propensity
    loss.backward()
    optimizer.step()
```

2.5 *Evaluation*

We evaluate DragonNet [1] by comparing predicted treatment effects to true effects.

Listing 17.6 Evaluating DragonNet performance

```python
with torch.no_grad():
    y0_pred, _ = model(torch.tensor(X_test, dtype=torch.
        float32), torch.zeros((len(X_test),1)))
    y1_pred, _ = model(torch.tensor(X_test, dtype=torch.
        float32), torch.ones((len(X_test),1)))

pred_effect = (y1_pred - y0_pred).squeeze().numpy()
true_effect = np.full_like(pred_effect, Treatment_effect)

# PEHE
pehe = np.sqrt(np.mean((pred_effect - true_effect)**2))

# ATE Error
true_ate = np.mean(true_effect)
pred_ate = np.mean(pred_effect)
ate_error = np.abs(pred_ate - true_ate)

print(f"PEHE: {pehe:.4f}")
print(f"ATE Error: {ate_error:.4f}")
```

2.6 Interpretation

If DragonNet [1] successfully learns the treatment effect, we expect a low PEHE and a small ATE error, indicating it adjusted adequately for motivation and GPA when estimating the impact of online learning on academic performance.

Causal Inference for Educational Interventions

- **Evaluating Educational Programs**: Machine learning can be used to rigorously evaluate the effectiveness of educational interventions, such as online learning programs.
- **Propensity Scores in ML**: DragonNet's propensity score [2] estimation is analogous to predicting the likelihood of an event in machine learning, and it helps to balance student groups for a fair comparison.
- **Policy Decisions in Education**: The results of such causal analyses can inform policy decisions about resource allocation and program design in educational settings.

3 Case Study 3: Estimating the Effect of Advertising on Sales

3.1 Problem Description

A company wants to estimate the causal effect of spending on digital advertising campaigns on product sales.

However, high-spending campaigns might also be correlated with factors like product quality or seasonality, which also impact sales. Thus, we must control for confounders such as product rating and time of year.

3.2 Causal Graph (DAG)

The causal structure can be represented as (Fig. 3):

3.3 Simulating Data

We simulate a dataset following this causal structure:

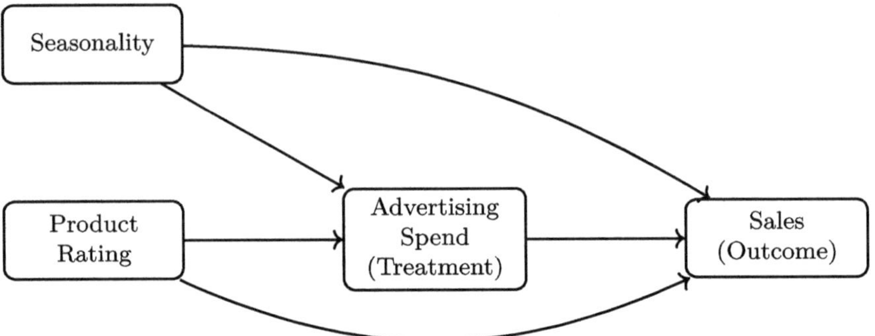

Fig. 3 Causal diagram: effect of advertising spend on sales, with seasonality and product rating as confounders

Listing 17.7 Simulating advertising and sales data

```python
import numpy as np

np.random.seed(2024)
n = 2500
Seasonality = np.random.choice([0, 1], size=n, p=[0.7, 0.3])
        # 0 = regular, 1 = holiday season
Rating = np.random.uniform(1.0, 5.0, size=n)  # Product
    ratings

# Propensity to spend more on ads
logit_p = 0.6 * Seasonality + 0.5 * (Rating - 3.0)
p = 1 / (1 + np.exp(-logit_p))
AdSpend = np.random.binomial(1, p)

# Sales outcome
Sales_baseline = 200 + 30 * Seasonality + 50 * (Rating -
    3.0) + np.random.normal(0, 20, n)
Treatment_effect = 40  # Advertising increases sales by 40
    units
Sales = Sales_baseline + AdSpend * Treatment_effect

# Assemble into features
X = np.stack([Seasonality, Rating], axis=1)
```

3.4 Modeling with CFRNet

We train a CFRNet model to predict the effect of advertising spend on sales.

Listing 17.8 Training CFRNet on advertising data

```python
# Split data
from sklearn.model_selection import train_test_split
```

```python
X_train, X_test, T_train, T_test, Y_train, Y_test =
    train_test_split(X, AdSpend, Sales, test_size=0.3)

# Convert to tensors
import torch
X_train_tensor = torch.tensor(X_train, dtype=torch.float32)
T_train_tensor = torch.tensor(T_train, dtype=torch.float32).
    unsqueeze(1)
Y_train_tensor = torch.tensor(Y_train, dtype=torch.float32).
    unsqueeze(1)

# Train CFRNet (reusing earlier CFRNet class)
model = CFRNet(input_dim=2)
optimizer = torch.optim.Adam(model.parameters(), lr=0.01)
criterion = torch.nn.MSELoss()

for epoch in range(100):
    optimizer.zero_grad()
    pred = model(X_train_tensor, T_train_tensor)
    loss = criterion(pred, Y_train_tensor)
    loss.backward()
    optimizer.step()
```

3.5 Evaluation

We evaluate CFRNet by comparing predicted treatment effects to true effects.

Listing 17.9 Evaluating CFRNet performance

```python
with torch.no_grad():
    y0_pred = model(torch.tensor(X_test, dtype=torch.float32
        ), torch.zeros((len(X_test),1)))
    y1_pred = model(torch.tensor(X_test, dtype=torch.float32
        ), torch.ones((len(X_test),1)))

pred_effect = (y1_pred - y0_pred).squeeze().numpy()
true_effect = np.full_like(pred_effect, Treatment_effect)

# PEHE
pehe = np.sqrt(np.mean((pred_effect - true_effect)**2))

# ATE Error
true_ate = np.mean(true_effect)
pred_ate = np.mean(pred_effect)
ate_error = np.abs(pred_ate - true_ate)

print(f"PEHE: {pehe:.4f}")
print(f"ATE Error: {ate_error:.4f}")
```

3.6 Interpretation

If the CFRNet model is trained correctly, it should estimate the increase in sales due to advertising spending with low PEHE and a small ATE error. Adjusting for seasonality and product ratings is crucial to obtain unbiased causal estimates in this setting.

Causal Marketing Analytics with ML

- **Marketing Campaign Effectiveness**: Machine learning models can estimate the causal impact of marketing campaigns (e.g., advertising spend) on sales, going beyond simple correlation analysis.
- **Attribution Modeling**: Causal inference techniques can contribute to more accurate attribution modeling, determining how each marketing channel contributes to conversions.
- **Optimizing Marketing Spend**: Businesses can optimize their spending and maximize ROI by understanding the causal effects of different marketing strategies.

4 Case Study 4: Estimating the Effect of a New Medication on Patient Recovery

4.1 Problem Description

Suppose we want to estimate the causal effect of administering a new medication on patient recovery time after surgery.

However, healthier patients may be more likely to receive the medication early, leading to confounding. Thus, we must adjust for pre-surgery health status and age.

4.2 Causal Graph (DAG)

The causal structure can be represented as (Fig. 4):

4.3 Simulating Data

We simulate a dataset following this causal structure:

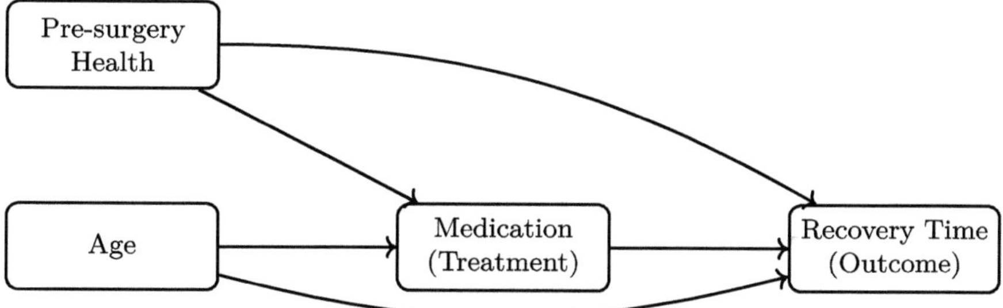

Fig. 4 Causal diagram: effect of medication on recovery time, with age and pre-surgery health as confounders

Listing 17.10 Simulating medication and recovery data

```python
import numpy as np

np.random.seed(2025)
n = 2000
Health = np.random.normal(0, 1, size=n)   # Health score
Age = np.random.normal(50, 10, size=n)    # Age in years

# Propensity to receive medication
logit_p = 0.8 * Health - 0.03 * Age
p = 1 / (1 + np.exp(-logit_p))
Medication = np.random.binomial(1, p)

# Recovery time outcome (lower is better)
Recovery_baseline = 10 - 2 * Health + 0.05 * Age + np.random
    .normal(0, 1, size=n)
Treatment_effect = -1.5   # Medication reduces recovery time
    by 1.5 days
RecoveryTime = Recovery_baseline + Medication *
    Treatment_effect

# Assemble into features
X = np.stack([Health, Age], axis=1)
```

4.4 Modeling with DragonNet

We train a DragonNet [1]model to predict the effect of medication on recovery time.

Listing 17.11 Training DragonNet on medication data

```python
# Split data
from sklearn.model_selection import train_test_split
X_train, X_test, T_train, T_test, Y_train, Y_test =
    train_test_split(X, Medication, RecoveryTime, test_size
    =0.3)
```

```python
# Convert to tensors
import torch
X_train_tensor = torch.tensor(X_train, dtype=torch.float32)
T_train_tensor = torch.tensor(T_train, dtype=torch.float32).
    unsqueeze(1)
Y_train_tensor = torch.tensor(Y_train, dtype=torch.float32).
    unsqueeze(1)

# Train DragonNet (reusing earlier DragonNet class)
model = DragonNet(input_dim=2)
optimizer = torch.optim.Adam(model.parameters(), lr=0.01)
criterion_outcome = torch.nn.MSELoss()
criterion_propensity = torch.nn.BCELoss()

for epoch in range(100):
    optimizer.zero_grad()
    y_pred, t_pred = model(X_train_tensor, T_train_tensor)
    loss_outcome = criterion_outcome(y_pred, Y_train_tensor)
    loss_propensity = criterion_propensity(t_pred,
        T_train_tensor)
    loss = loss_outcome + loss_propensity
    loss.backward()
    optimizer.step()
```

4.5 *Evaluation*

We evaluate DragonNet [1] by comparing predicted treatment effects to true effects.

Listing 17.12 Evaluating DragonNet performance

```python
with torch.no_grad():
    y0_pred, _ = model(torch.tensor(X_test, dtype=torch.
        float32), torch.zeros((len(X_test),1)))
    y1_pred, _ = model(torch.tensor(X_test, dtype=torch.
        float32), torch.ones((len(X_test),1)))

pred_effect = (y1_pred - y0_pred).squeeze().numpy()
true_effect = np.full_like(pred_effect, Treatment_effect)

# PEHE
pehe = np.sqrt(np.mean((pred_effect - true_effect)**2))

# ATE Error
true_ate = np.mean(true_effect)
pred_ate = np.mean(pred_effect)
ate_error = np.abs(pred_ate - true_ate)

print(f"PEHE: {pehe:.4f}")
print(f"ATE Error: {ate_error:.4f}")
```

4.6 Interpretation

If the DragonNet model is correctly trained, we expect low PEHE and ATE error values, indicating it successfully adjusted for health status and age when estimating the medication's effect on recovery time.

Causal Drug Discovery and Development

- **Drug Efficacy Estimation**: Machine learning can be used to estimate the causal effect of new medications on patient outcomes, even in observational studies where treatment assignment is not randomized.
- **Personalized Treatment Effects**: By modeling heterogeneous treatment effects, machine learning can help identify which patients are most likely to benefit from a particular medication.
- **Accelerating Clinical Trials**: Causal inference techniques can potentially accelerate clinical trials by providing more efficient ways to analyze data and identify promising drug candidates.

5 Mini Case Study 5: Effect of Remote Work on Employee Productivity

5.1 Problem Description

A company wants to estimate whether allowing employees to work remotely increases their productivity.

However, more disciplined employees may choose remote work, creating confounding.

5.2 Causal Graph (DAG)

See Fig. 5.

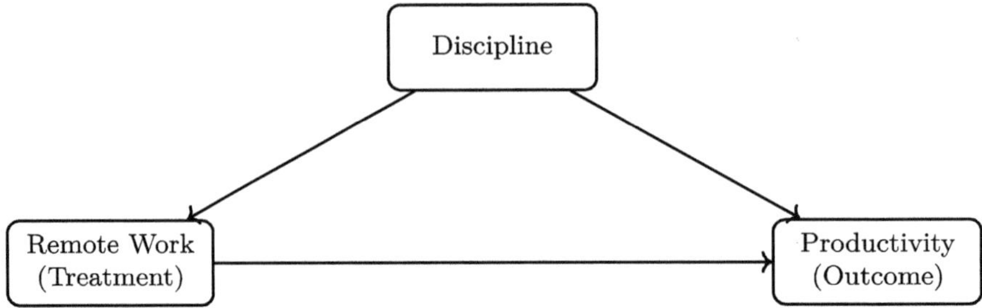

Fig. 5 Causal diagram: effect of remote work on productivity, with discipline as a confounder

5.3 *Simulation Setup*

We simulate a small dataset:

Listing 17.13 Simulating remote work data

```python
import numpy as np

np.random.seed(2026)
n = 1000
Discipline = np.random.normal(0, 1, size=n)

# Propensity to choose remote work
logit_p = 1.2 * Discipline
p = 1 / (1 + np.exp(-logit_p))
RemoteWork = np.random.binomial(1, p)

# Productivity outcome
Productivity_baseline = 50 + 10 * Discipline + np.random.normal(0, 5, n)
Treatment_effect = 3  # Remote work increases productivity by 3 units
Productivity = Productivity_baseline + RemoteWork * Treatment_effect

X = Discipline.reshape(-1, 1)
```

5.4 *Modeling Note*

A simple CFRNet [3] or DragonNet [1] model can be trained on this dataset to estimate the causal effect of remote work.

Here, the key confounder—discipline must be adjusted adequately for accurate inference.

Causal Analysis of Organizational Policies

- **Policy Impact Assessment**: Machine learning can be employed to assess the causal impact of organizational policies, such as remote work arrangements, on employee productivity and other relevant outcomes.
- **HR Analytics**: Causal inference techniques can enhance HR analytics by providing more accurate and reliable insights into the effectiveness of HR practices.
- **Data-Driven Management**: By understanding the causal effects of different policies, organizations can make more informed and data-driven management decisions.

6 Summary of Case Studies

Throughout this chapter, we explored several real-world causal inference [4] scenarios to consolidate the concepts introduced earlier in the book.

Each case study followed a systematic structure:

- **Problem Framing**: Identify a real-world causal question involving treatment and outcome.
- **Causal Graph (DAG) Construction**: Visualize confounders, treatment assignment mechanisms, and outcome relationships.
- **Data Simulation**: Create synthetic datasets reflecting the specified causal structure.
- **Modeling**: Apply counterfactual prediction models like CFRNet and DragonNet.
- **Evaluation**: Quantify model performance using PEHE and ATE error metrics.

Lessons Learned

- **Importance of Confounder Adjustment**: Across all case studies, failing to adjust for key confounders (like baseline health, prior GPA, motivation, product rating, or discipline) would have led to biased causal estimates.
- **Role of Causal Diagrams**: Directed acyclic graphs (DAGs) helped clarify assumptions about the data-generating process and informed modeling decisions.
- **Data Simulation Practice**: Simulating data reinforced understanding of how different variables interact causally and how confounding can affect treatment assignment and outcomes.
- **Modeling Techniques**: CFRNet and DragonNet demonstrated how neural networks can be adapted to learn counterfactual outcomes in observed confounders.
- **Evaluation Challenges**: Without access to true counterfactuals in real-world data, estimating model performance requires careful proxy metrics like PEHE and ATE error, often validated using synthetic data.

Key Takeaways

- Always begin causal analysis by drawing a causal graph.
- Understand which variables act as confounders, mediators, or colliders.
- Choose modeling strategies that align with the causal structure.
- Evaluating causal models requires thoughtful metric design, especially when counterfactuals are not directly observable.

These case studies are designed to illustrate causal inference principles and encourage readers to practice building causal models from scratch, thereby developing strong causal reasoning skills.

7 Exercises

1. **Instrument Selection Exercise** Imagine you want to estimate the causal effect of job training on employee wages. Suggest two possible instrumental variables [5] and explain why they may or may not satisfy the three main IV assumptions:

   - Relevance (instrument affects treatment)
   - Independence (instrument is independent of unmeasured confounders)
   - Exclusion restriction (instrument affects outcome only through treatment)

2. **Causal Graph Analysis**

   Consider the following causal graph:

   $$A \rightarrow B \rightarrow C, \quad A \rightarrow C$$

   (a) Is B a confounder, a mediator, or a collider in the relationship between A and C?
   (b) If we want to estimate the causal effect of A on C, should we adjust for B? Why or why not?

References

1. Shi, Chao, David, M., Blei, and Victor, Veitch. 2019. Adapting neural networks for the estimation of treatment effects, in *Advances in neural information processing systems (NeurIPS)*.
2. Rosenbaum, Paul, R., and Donald, B. Rubin. 1983. The central role of the propensity score in observational studies for causal effects, in Biometrika 70 (1): 41–55.
3. Johansson, Fredrik, D., Uri. Shalit, and David Sontag. 2016. Learning representations for counterfactual inference, in *Proceedings of the 33rd international conference on machine learning (ICML)*, 1386–1395.

4. Pearl, Judea. 2009. *Causality: Models, reasoning, and inference*. 2nd ed. Cambridge University Press.
5. Angrist, Joshua, D., Guido, W., Imbens, and Donald, B. Rubin. 1996. Identification of causal effects using instrumental variables. *Journal of the American Statistical Association* 91(434): 444–455.

Solutions to Exercises

1 Introduction: Causal Inference Exercises for Machine Learning

The exercises throughout this book are designed to reinforce your understanding of causal inference [1] concepts and their application. For machine learning engineers, these exercises provide opportunities to:

- **Think Causally About Data**: Practice identifying potential confounders, mediators, and other causal structures that can influence the data used to train machine learning models.
- **Design Causal-Aware Models**: Explore how causal graphs and other tools can guide the design of machine learning architectures that are more robust and interpretable.
- **Evaluate Model Limitations**: Critically assess when standard machine learning techniques might fail due to causal complexities and when causal inference methods are necessary.

The solutions in this chapter offer detailed explanations and insights to help you develop a strong foundation in causal thinking for machine learning.

2 Introduction to Causal Thinking

1. **Define causality in your own words. How is it different from correlation?**
   Causality means that a change in one variable directly results in a change in another. It implies a directional, cause-and-effect relationship. In contrast, correlation only indicates that two variables move together statistically, without implying one causes the other.

D. Rajamanickam, *Causal Inference for Machine Learning Engineers*,
https://doi.org/10.1007/978-3-031-99680-1_18

2. **Give a real-world example where confusing correlation with causation could lead to a wrong decision**.
   Suppose data shows that ice cream sales and drowning incidents are correlated. One might wrongly conclude that eating ice cream causes drowning. In reality, both are influenced by a third factor—hot weather—making this a case of spurious correlation.

3. **What is the primary goal of causal inference?**
   The main goal of causal inference is to estimate the effect of interventions or treatments by understanding how changing one variable (the cause) influences another (the effect), while accounting for confounding and ensuring the relationship is truly causal.

4. **Describe an everyday situation where counterfactual thinking helps you make better choices**.
   Imagine you failed a test and wonder, "What if I had studied more?" This counterfactual helps you realize the consequences of your choices and motivates better preparation for the next exam. Such thinking allows for learning from experience by imagining alternate outcomes.

3 Treatments, Outcomes, and Confounding: Core Concepts

1. **Define treatment, outcome, and confounder in the context of causal inference**.
   Treatment: A variable representing an intervention or action (e.g., drug administering).
   Outcome: The variable of interest that may be influenced by the treatment (e.g., recovery status).
   Confounder: A variable that affects both the treatment and the outcome, creating a spurious association if not controlled (e.g., age influencing treatment choice and recovery rate).

2. **Provide an example of a confounding variable and explain how it affects causal conclusions**. In studying whether exercise (treatment) improves mental health (outcome), socioeconomic status could be a confounder—it influences both the likelihood of exercising and access to mental health resources. If not adjusted for, one might overestimate the effect of exercise alone.

3. **Why is adjusting for confounders critical when estimating treatment effects?**
   Adjusting for confounders removes bias due to backdoor paths in the causal graph. It ensures that the estimated effect of the treatment on the outcome is not influenced by variables that affect both.

4. **What happens if we adjust for a collider instead of a confounder?**
   Conditioning on a collider (a variable influenced by two or more variables in the graph) opens a non-causal path between them, introducing bias. For example, if skill and luck influence hiring, conditioning on hiring status creates a spurious negative correlation between skill and luck.

4 Causal Estimation Basics

1. **Explain what an Average Treatment Effect (ATE) is.**
   The Average Treatment Effect (ATE) is the expected outcome difference between a population that receives the treatment and one that does not. Formally, it is defined as:
   $$\text{ATE} = \mathbb{E}[Y(1) - Y(0)],$$

   where $Y(1)$ is the potential outcome under treatment and $Y(0)$ is the potential outcome under control.

2. **Define Individual Treatment Effect (ITE) and how it differs from ATE.**
   The Individual Treatment Effect (ITE) is the difference in potential outcomes for a specific individual:
   $$\text{ITE}_i = Y_i(1) - Y_i(0).$$

   While ATE is an average over a population, ITE captures the treatment effect for a specific person and reflects treatment heterogeneity.

3. **Why can't we directly observe an individual's ITE in practice?**
   We cannot observe both potential outcomes $Y_i(1)$ and $Y_i(0)$ for the same individual, because in reality, we only observe the outcome under the treatment received. The counterfactual outcome remains unobserved.

4. **Give an example where estimating the ATE is insufficient for decision-making.**
   In personalized medicine, a drug may help some patients and harm others. If the ATE is close to zero, it hides this variability. A doctor needs individual-level estimates (ITE) to make tailored treatment decisions.

5 Causal Graphs: Structure and Assumptions

1. **Draw a simple Directed Acyclic Graph (DAG) showing confounding.**
   Below is a simple DAG where variable Z is a confounder of the relationship between treatment T and outcome Y:

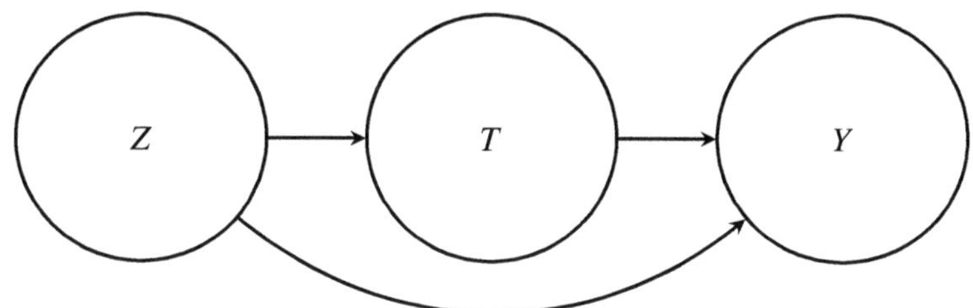

2. **What does it mean for a graph to be acyclic?**
A graph is acyclic if it has no directed cycles. That means there is no path following the direction of arrows that starts and ends at the same node. In causal inference, this ensures variables cannot cause themselves directly or indirectly.
3. **Explain the role of backdoor paths in causal inference.**
Backdoor paths are non-causal paths that introduce spurious associations between treatment and outcome due to common causes (confounders). If not blocked, these paths bias the estimate of the treatment effect.
4. **How does blocking backdoor paths help in estimating causal effects?**
Blocking backdoor paths (usually by conditioning on confounders) eliminates spurious associations. This isolates the direct causal effect of the treatment on the outcome, leading to more accurate and unbiased estimates.

6 Interventions and Counterfactuals

1. **Explain the concept of an intervention using the do-operator.**
An intervention, denoted by the do-operator $do(T = t)$, represents an external action that forces the treatment variable T to take a specific value t, regardless of its natural causes. It simulates a controlled experiment where we break the usual causal links into T and examine the resulting distribution of the outcome.
2. **What is the difference between $P(Y \mid T)$ and $P(Y \mid do(T))$?**
$- P(Y \mid T)$: The observational distribution; it includes both causal and non-causal associations (e.g., confounding).
$- P(Y \mid do(T))$: The interventional distribution; it reflects the causal effect of T on Y by simulating an intervention that severs incoming edges into T.
In general, $P(Y \mid T) \neq P(Y \mid do(T))$ when confounding is present.
3. **Define a counterfactual outcome and explain why it is important.**
A counterfactual outcome represents what would have happened to an individual under a different treatment than the one actually received. For instance, if a person took a drug and recovered, the counterfactual asks: "Would they have recovered without the drug?" Counterfactuals are central to personalized decision-making, fairness evaluation, and understanding causality at the individual level.
4. **Give a practical example where counterfactual reasoning changes a decision.**
Consider a student who failed an exam and asks, "Would I have passed if I had studied more?" This counterfactual insight may guide them to change their study habits. In a business setting, a company may ask, "Would sales have increased if we had launched the ad campaign?" Such reasoning is key for learning from past decisions and improving future actions.

7 Introduction to Do-Calculus

1. **What is do-calculus and why is it important in causal inference?**
   Do-calculus [1] is a set of formal rules developed by Judea Pearl [1] that allow us
   to transform expressions involving interventions (e.g., $P(Y \mid do(T))$) into expressions involving only observational quantities (e.g., $P(Y \mid T, Z)$)—under certain
   graphical conditions. It is essential for determining when and how causal effects
   can be identified from observational data, especially in the presence of unmeasured
   confounding or complex causal structures.
2. **Explain the intuition behind one of the rules of do-calculus**.
   Rule 1 (Insertion/deletion of observations): If a variable Z is independent of
   the outcome Y given treatment T and context X in the graph where incoming
   arrows to T are removed, then conditioning on Z doesn't affect the causal effect:

$$P(Y \mid do(T), Z, X) = P(Y \mid do(T), X).$$

   Intuition: If Z is irrelevant to Y after intervening on T, adding or removing it
   doesn't change anything.
3. **Provide an example where do-calculus[1] allows identification of a causal
   effect**.
   Suppose we cannot directly estimate $P(Y \mid do(T))$ due to unobserved confounding between T and Y, but a mediator M lies on the path from T to Y, and there
   are no backdoor paths from M to Y. Using do-calculus[1], we may derive:

$$P(Y \mid do(T)) = \sum_m P(Y \mid M = m, T) P(M = m \mid do(T)).$$

   This expression rewrites the interventional distribution using observed data and
   graphical assumptions.
4. **Why do we need formal rules for moving between observational and interventional distributions?**
   Observational data often contains confounding, making it unreliable for estimating causal effects directly. Formal rules like those in do-calculus [1] provide
   a principled way to determine when and how interventional quantities can be
   computed from observational data, ensuring the validity of causal claims.

8 Backdoor and Frontdoor Criteria

1. **What is the confounding challenge in causal inference, and why does it arise
   in observational data?**
   Confounding occurs when a variable influences the treatment and the outcome,
   creating a spurious association between them. In observational data, treatments are

not randomly assigned, so such confounders are often present, making it difficult to determine whether an observed association reflects a true causal effect.

2. **Define the backdoor criterion. What conditions must a set of variables satisfy to meet this criterion?**
A set of variables Z satisfies the backdoor criterion with respect to treatment T and outcome Y if:

   (a) No element of Z is a descendant of T, and
   (b) Z blocks all backdoor paths from T to Y (paths that end with an arrow into T).

   If such a Z exists, it can be used to adjust for confounding.

3. **Explain the backdoor adjustment formula. How does it allow us to estimate causal effects in the presence of confounding?**
The backdoor adjustment formula is:

$$P(Y \mid do(T)) = \sum_z P(Y \mid T, Z = z)P(Z = z).$$

   By conditioning on the appropriate set Z, we block all backdoor paths, thereby removing confounding bias and enabling valid estimation of the causal effect from observational data.

4. **Define the frontdoor criterion. What are the key differences between the frontdoor and backdoor criteria?**
The frontdoor criterion applies when confounding between T and Y cannot be removed by adjusting for observed covariates. It relies on an intermediate variable M (a mediator) that lies on the path from T to Y, and requires:

   - M is affected by T,
   - Y is affected by M,
   - All backdoor paths from T to M, and from M to Y, are blocked by observed variables.

   Unlike the backdoor criterion [1], frontdoor criterion [1] does not require direct observation of confounders between T and Y, as long as the mediator provides a sufficient causal channel.

5. **Under what conditions can the frontdoor criterion be used to estimate causal effects?**
The frontdoor criterion can be used when:

   (a) The treatment T affects the mediator M,
   (b) The mediator M affects the outcome Y,
   (c) All backdoor paths from T to M, and from M to Y, are blocked by observed variables, and
   (d) There are no unblocked paths from T to Y that do not go through M.

   When these conditions are met, causal effects can be identified even in unmeasured confounding between T and Y.

9 Advanced Causal Inference Methods

- **Instrumental Variables**: Generate synthetic data violating IV assumptions (e.g., instrument directly affects the outcome) and analyze how the violation impacts the 2SLS estimates. Illustrate the violation with a DAG.

Approach:

- Simulate data where instrument Z affects treatment T, but also directly affects outcome Y, violating the exclusion restriction.
- Use 2SLS to estimate the effect of T on Y, assuming Z is a valid instrument.
- Compare the estimate to the ground-truth causal effect to observe bias.

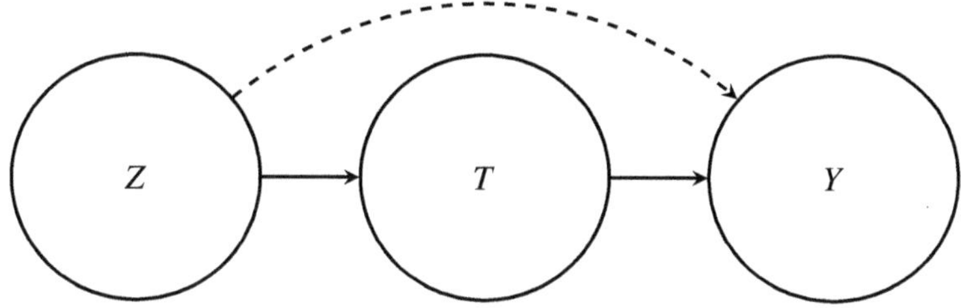

DAG

Insight: The dashed edge from $Z \rightarrow Y$ violates the exclusion restriction, leading to biased 2SLS estimates. This highlights the importance of validating IV assumptions before use.

1. **Mediation Analysis**: Conduct a mediation analysis with multiple mediators (e.g., job training $\rightarrow$ increased confidence $\rightarrow$ increased job search $\rightarrow$ better job $\rightarrow$ higher earnings).

   Approach:

   - Define path model with sequential mediators M_1, M_2, M_3, M_4 where:

   $$T \rightarrow M_1 \rightarrow M_2 \rightarrow M_3 \rightarrow M_4 \rightarrow Y$$

   - Use structural equation modeling or sequential regressions to estimate direct and indirect effects.
   - Apply the product of coefficients method to compute mediation effects.

   Insight: Mediation helps decompose the total effect of job training on earnings into actionable causal chains. It also clarifies which intermediate factors contribute most to the outcome.

2. **Causal Discovery**: Compare the performance of PC and FCI algorithms using simulated data with a known latent confounder.

Approach:

- Simulate a dataset with 5 observed variables and 1 hidden confounder.
- Apply the PC algorithm (assumes no latent confounding) and the FCI algorithm (accounts for latent variables).
- Compare the output PAGs (Partial Ancestral Graphs) to the ground-truth structure.

Insight: PC may produce incorrect edges due to hidden confounding, while FCI can detect latent structures via bi-directed edges or partially oriented edges, improving robustness in real-world scenarios.

3. **Transportability**: Simulate datasets reflecting two populations with different distributions of a key covariate. Apply the transportability formula to adjust causal effect estimates and visualize the distributions and the adjustment process.

   Approach:

   - Simulate source population $\mathcal{P}_1$ and target population $\mathcal{P}_2$, differing in a covariate Z.
   - Estimate $P(Y \mid do(T), Z)$ in $\mathcal{P}_1$, and use reweighting with $P(Z)$ from $\mathcal{P}_2$ to obtain the transported estimate:

   $$P^{\mathcal{P}_2}(Y \mid do(T)) = \sum_z P^{\mathcal{P}_1}(Y \mid do(T), z) \cdot P^{\mathcal{P}_2}(z)$$

   - Visualize the distributions of Z in both populations and how the reweighted average shifts the causal estimate.

   Insight: Transportability allows the reuse of experimental data from one population to infer effects in another. This is critical in fields like medicine, where conducting trials in every subgroup may not be feasible.

10 Causal Inference Meets Deep Learning

1. **Why is deep learning useful for causal inference in high-dimensional settings?**
   Deep learning is effective in high-dimensional settings because it can learn complex, nonlinear relationships between variables. In causal inference, this helps:

   - Extract relevant features from high-dimensional covariates (e.g., text, images),
   - Model intricate interactions between treatment, covariates, and outcomes,
   - Balance treated and control groups by learning representations that reduce bias.

2. **Explain the key challenges in applying deep learning to causal problems**.
   Key challenges include:

   - **Unobserved counterfactuals**: We can't observe both potential outcomes for each individual.
   - **Treatment assignment bias**: Due to confounding, deep models may pick up on spurious patterns.
   - **Lack of identifiability**: Neural networks can fit data well but may not reflect correct causal relationships.
   - **Interpretability**: Deep models often act as black boxes, which conflicts with the transparency required in causal inference.

3. **Describe the purpose of representation learning in causal models**.
   Representation learning in causal inference aims to learn latent features of covariates that satisfy balance or ignorability conditions. Good representations:

   - Remove confounding bias,
   - Make treated and control units comparable,
   - Improve generalization across domains and reduce covariate shift.

4. **What could go wrong if deep learning models ignore treatment assignment bias?**
   If treatment assignment bias is ignored, models may:

   - Learn patterns based on spurious correlations,
   - Confuse predictors of treatment with causal predictors of outcome,
   - Produce biased or misleading treatment effect estimates,
   - Fail to generalize across populations or environments.

This leads to incorrect inferences and poor decision-making.

11 Simulating Causal Data and Evaluation Metrics

1. **Why is simulation necessary in causal machine learning?**
   Simulation is necessary because, in real-world observational data, we never observe both potential outcomes $Y(0)$ and $Y(1)$ for the same individual. Without this, we can't directly evaluate the accuracy of Individual Treatment Effect (ITE) predictions. Simulated datasets allow us to generate ground-truth counterfactuals, enabling proper benchmarking and model comparison.
2. **Describe how you would simulate treatment assignment in a synthetic dataset**.
   To simulate treatment assignment:

   - Generate covariates $X \sim P(X)$ (e.g., from a multivariate normal distribution).

- Define a propensity score model, e.g., $e(X) = P(T = 1 \mid X) = \text{sigmoid}(w^\top X)$.
- Sample treatment $T \sim \text{Bernoulli}(e(X))$.
- Define outcome models for $Y(0)$ and $Y(1)$, then use the actual treatment to assign $Y = Y(T)$.

3. **Define PEHE and ATE Error**.
   – **PEHE (Precision in Estimation of Heterogeneous Effect)**:

$$\text{PEHE} = \sqrt{\mathbb{E}\left[(I\hat{T}E_i - ITE_i)^2\right]},$$

where $I\hat{T}E_i$ is the model's predicted ITE and $ITE_i = Y_i(1) - Y_i(0)$ is the true effect.
   – **ATE Error**:

$$\text{ATE Error} = \left| A\hat{T}E - \text{ATE}_{\text{true}} \right|,$$

which measures the absolute error between the model's estimated average treatment effect and the true value.

4. **Explain how simulated data helps in model evaluation**.
   Simulated data provides known ground-truth potential outcomes $Y(0), Y(1)$, enabling direct computation of:

- ITE estimation accuracy (via PEHE),
- ATE accuracy,
- Evaluation of how well models generalize across different treatment assignment mechanisms and outcome surfaces.

It allows controlled experimentation, stress-testing of models, and debugging of assumptions in a reproducible environment.

12 Balancing Representations with Causal Deep Learning (CFRNet)

1. **What is the main goal of balancing representations in causal inference?**
   The goal of balancing representations is to ensure that the distributions of covariates for treated and control groups are similar in the learned representation space. This helps to:

- Reduce confounding bias,
- Make comparisons between treated and control units fair,
- Improve the estimation of individual and average treatment effects.

2. **Explain how CFRNet extends TARNet**.
   CFRNet (Counterfactual Regression Network) [2] extends TARNet (Treatment-Agnostic Representation Network) [3] by:

   - Adding a loss term that penalizes imbalance between treated and control groups in the representation space,
   - Specifically incorporating an Integral Probability Metric (IPM) into the objective to encourage balanced distributions.

   TARNet [3] learns shared representations and separate heads for outcomes, but CFRNet [2] actively enforces similarity between groups.
3. **What is an Integral Probability Metric (IPM) and how is it used in CFRNet?**
   The IPM measures the distance between two probability distributions over the representation space. Formally, it is:

$$\mathrm{IPM}(\mathcal{F}, P, Q) = \sup_{f \in \mathcal{F}} \left| \mathbb{E}_P[f(X)] - \mathbb{E}_Q[f(X)] \right|.$$

   In CFRNet, IPM penalizes discrepancies between the representations of treated and control groups. A smaller IPM means a better balance.
4. **Why is balancing important even if we have high prediction accuracy?**
   High prediction accuracy does not imply valid causal inference. A model could predict outcomes well due to correlations rather than causal structure. If the treated and control groups are imbalanced, the model may extrapolate incorrectly when estimating counterfactuals. Balancing ensures the model generalizes appropriately and yields unbiased treatment effect estimates.

13 Propensity Scores in Causal Deep Learning

1. **What role does the propensity score play in causal inference?**
   The propensity score is the probability of receiving treatment given observed covariates:
$$e(X) = P(T = 1 \mid X).$$

   It helps reduce bias from confounding by summarizing covariates into a single scalar. Propensity scores match, stratify, or weight samples so that treated and control groups become comparable, simulating randomized experiments.
2. **How does DragonNet incorporate the propensity score into its architecture?**
   DragonNet extends CFRNet by including a third output head that explicitly estimates the propensity score alongside the outcome heads. This multi-task structure allows the network to:

   - Simultaneously learn outcome predictions and treatment assignment probabilities,

- Use the estimated propensity score as part of the loss function to encourage balance and improve generalization.

3. **Explain targeted regularization and its purpose in DragonNet.**
   Targeted regularization in DragonNet refers to adding a loss term that encourages the learned representations to align with a statistical target—in this case, better balancing via the propensity score. It serves to:

   - Fine-tune predictions using prior knowledge from propensity scores,
   - Improve stability of treatment effect estimation by constraining the learning space.

4. **What are the challenges of balancing prediction and propensity loss during training?**
   The main challenge is multi-objective optimization. The outcome loss and propensity loss may:

   - Compete, leading to unstable gradients,
   - Require careful tuning of weighting factors,
   - Cause overfitting to one component (e.g., perfect treatment prediction but poor counterfactual generalization),
   - Need early stopping or dynamic learning rates to stabilize training.

   Achieving the right trade-off is critical for accurate and unbiased causal inference.

14   Evaluating Causal Models Without Counterfactuals

1. **Why is evaluating causal models more difficult than evaluating standard supervised models?**
   In standard supervised learning, we observe true labels for every instance, enabling direct comparison with predictions. In causal inference, however, we only observe one potential outcome for each unit—either $Y(0)$ or $Y(1)$, but never both. This missing data problem makes it difficult to evaluate the accuracy of estimated treatment effects, particularly the Individual Treatment Effect (ITE).

2. **Define PEHE, ATE Error, and Policy Risk.**

   - **PEHE (Precision in Estimation of Heterogeneous Effect):**

   $$\text{PEHE} = \sqrt{\mathbb{E}\left[(\hat{ITE}_i - ITE_i)^2\right]}.$$

   It measures the accuracy of the estimation of individual treatment effect.

   - **ATE Error:**
   $$\text{ATE Error} = \left|\hat{ATE} - \text{ATE}_{\text{true}}\right|.$$

It quantifies the difference between the estimated and true average treatment effect.

- **Policy Risk**:

$$\text{Policy Risk} = \mathbb{E}\left[Y_i^{\pi_0(X_i)} - Y_i^{\pi(X_i)} \right],$$

where $\pi(X)$ is the learned treatment policy and $\pi_0(X)$ is the optimal policy. It evaluates how well a learned policy performs compared to the best decision rule.

3. **How can simulated datasets help evaluate causal models?**
   Simulated datasets provide full knowledge of the data-generating process, including both potential outcomes $Y(0)$ and $Y(1)$ for each individual. This allows direct computation of metrics like PEHE and ATE Error, enabling rigorous benchmarking, model comparison, and stress-testing under controlled conditions.
4. **Describe two strategies for evaluating causal models when counterfactuals are unobserved.**

   - **Use of real-world semi-synthetic benchmarks**: Replace the real outcomes in a real dataset with simulated outcomes (e.g., using a known data-generating process) so that ground truth is available for evaluation.
   - **Proxy or indirect validation**: Evaluate model performance on downstream tasks, such as policy risk, uplift modeling, or consistency with known randomized trial results. Another approach is sensitivity analysis, which assesses the robustness of assumptions.

15 Advanced Topics in Causal Inference

1. **Causal Discovery**

   (a) Purely observational data cannot uniquely determine the true causal graph without additional assumptions because multiple graphs may encode the same conditional independence relationships (i.e., be Markov equivalent). Assumptions like faithfulness, acyclicity, and no hidden confounding are required to identify a unique graph.
   (b) The PC (Peter-Clark) algorithm has two phases:

   - **Skeleton identification**: Start with a fully connected undirected graph and remove edges based on conditional independence tests.
   - **Orientation phase**: Apply a set of orientation rules (e.g., collider detection) to direct as many edges as possible without introducing cycles.

2. **Instrumental Variables**

   (a) The three key assumptions for a valid instrumental variable (IV) are:

- **Relevance**: The instrument Z must be correlated with the treatment T (i.e., $Z \to T$).
- **Exclusion Restriction**: The instrument affects the outcome Y only through T, not directly (i.e., no $Z \to Y$ path except through T).
- **Independence**: The instrument Z is independent of any unmeasured confounders affecting T and Y.

(b) A possible IV could be the distance to the nearest gym. It likely influences the likelihood of exercising (relevance). Still, it does not directly affect weight loss (exclusion restriction), assuming people are not inherently healthier just because they live closer to a gym (independence).

3. **Mediation Analysis**

(a)
- **Natural Direct Effect (NDE)**: The effect of treatment on the outcome not through the mediator (i.e., holding the mediator constant at its natural level).
- **Natural Indirect Effect (NIE)**: The portion of the treatment effect that operates through the mediator.

(b) **Causal Diagram**:

$$\text{Diet} \to \text{Weight Change} \to \text{Blood Pressure}$$

To estimate the direct and indirect effects:

- Fit a model for the mediator (e.g., weight change) given treatment.
- Fit an outcome model (e.g., blood pressure) based on treatment and the mediator.
- Use the models to compute counterfactual outcomes for the same treatment level but different mediator states to isolate NDE and NIE.

4. **Sensitivity Analysis**

(a) In Rosenbaum bounds, the parameter Γ quantifies the degree of hidden bias due to unmeasured confounding. Specifically, it represents how much the odds of treatment assignment could differ between treated and control units with the same observed covariates due to hidden variables.

(b) If your study is robust up to $\Gamma = 2.0$, this implies that an unmeasured confounder would need to double the odds of treatment assignment to nullify your causal conclusion. The higher the Γ, the more robust your result.

5. **Critical Thinking**

(a) An IV could be a random invitation to the online program (e.g., lottery or staggered rollout). This assignment mechanism affects participation (treatment) but not test scores directly, and is independent of motivation.

(b) Sensitivity analysis can be used to examine how strong an unmeasured confounder must be to change the observed effect estimate. Methods like Rosenbaum bounds or bounding approaches can provide thresholds under which the causal effect would no longer be statistically significant.

References

1. Pearl, Judea. 2009. *Causality: Models, reasoning, and inference*. 2nd ed.. Cambridge University Press.
2. Johansson, Fredrik, D., Uri, Shalit, and David, Sontag. 2016. Learning representations for counterfactual inference, In *Proceedings of the 33rd international conference on machine learning (ICML)*, 1386–1395.
3. Shalit, Uri, Fredrik, D., Johansson, and David, Sontag. 2017. Estimating individual treatment effect: Generalization bounds and algorithms. In *Proceedings of the 34th international conference on machine learning (ICML)*, 3076–3085.

Index

If you have any concerns about our products,
you can contact us on
ProductSafety@springernature.com

In case Publisher is established outside the EU,
the EU authorized representative is:
Springer Nature Customer Service Center GmbH
Europaplatz 3, 69115 Heidelberg, Germany

Printed by Libri Plureos GmbH
in Hamburg, Germany